PRINCIPLES OF PHASE DIAGRAMS
IN MATERIALS SYSTEMS

PRINCIPLES OF PHASE DIAGRAMS IN MATERIALS SYSTEMS

PAUL GORDON

Professor and Chairman
Department of Metallurgical Engineering
Illinois Institute of Technology

McGraw-Hill Book Company

New York
St. Louis
San Francisco
Toronto
London
Sydney

**PRINCIPLES OF PHASE DIAGRAMS
IN MATERIALS SYSTEMS**

Library of Congress Catalog Card Number 68-11604

23793

1234567890 MAMM 7432106987

PREFACE

In the course of several years of teaching the concepts involved in phase transformations to student scientists and engineers, two impressions have gradually grown on me, namely, the beauty and importance of the thermodynamic method in phase transitions and the difficulty many students have in gaining a genuine understanding of the method. I have also become convinced that, at least for metallurgists and other materials scientists and engineers, a major reason for the latter difficulty lies in the absence of a text with the appropriate tone for such students. In the treatment of phase diagrams in particular, the available books appear to fall into one of the following categories: (1) they are heavily oriented toward the viewpoint of the chemist, and, therefore, though they treat the systems involved thermodynamically, they give the area of materials systems little attention; (2) they are highly mathematical, with a language and symbolism outside the realm of easy familiarity to the materials engineer; (3) they are treatises rather than texts; (4) they deal only briefly with phase diagrams as one of many subjects in a book of much more general nature; or (5) they give only what might be called the geometrical and phenomenological approach to phase diagrams, with little or no discussion of the thermodynamic foundations. I hasten to point out that many such books are excellently written and present viewpoints of vital interest and usefulness. Nevertheless, it has seemed to me that in the materials area there is a need for a relatively brief text limited to the subject of phase diagrams, adopting the thermodynamic viewpoint, and developed with the needs and interests of the materials student in mind. The present book has been written in an effort to meet these criteria. It has been aimed at the materials student in the upper undergraduate and first-year graduate levels.

The book is intended to cover primarily, but not exclusively, the *equilibrium* relationships in materials systems, emphasizing the thermodynamic foundations of equilibrium phase diagrams and the meanings of the lines in the diagrams: for example, a liquidus line is not merely the line above which all alloys are completely liquid, but it is the locus of temperatures at which liquids of various compositions become saturated with respect to certain solids during cooling, or it is the locus of liquid compositions which can be in equilibrium with certain solids. The thermodynamic origin of the equilibrium lines is described in terms of the free-energy criterion. In order to keep the mathematical symbolism relatively simple, the quantities fugacity and activity, as well as the activity coefficient, useful as they may be, have been omitted. I have found that within the scope of this book, these quantities are unnecessary and often lead to confusion rather than to clarification in the beginning student's mind.

Since one of my purposes in writing this book was to provide a short text on a limited subject which could be fitted into a course, or series of courses, at the discretion of the instructor, the scope has been purposely kept narrow. The kinetics of transformations and departures from equilibrium have been treated, but generally only to illustrate the manner in which the equilibrium relationships can be used as a basis for understanding the more complex nonequilibrium phenomena. Also the usual type of phenomenological discussions of cooling and heating transformations, though included, have been kept to a minimum. Furthermore, ternary and multicomponent diagrams have been only briefly mentioned. I have considered that the amount of space necessary to give such diagrams a reasonably complete treatment was unwarranted in this book because the thermodynamic principles involved could be more easily discussed for simple systems and because the geometrical treatment has been quite thoroughly covered in many other books. On the other hand, the effect of high pressure as a third thermodynamic variable (in addition to temperature and composition) in materials systems has been given more than the usual attention because of the increased experimental and practical interest in high-pressure treatment of liquids and solids.

In essence, then, this book consists of a discussion of the principles of equilibrium-phase-diagram relationships in one- and two-component materials systems. The underlying thermodynamic foundations of the diagrams are developed on the basis of the free-energy criterion, considering temperature, composition, and pressure as the relevant variables. The quasi-chemical and other atomistic viewpoints are introduced where they appear to lend clarification. Some kinetic and nonequilibrium concepts and some phenomenological discussions are included in sufficient detail so that the book will serve as a self-contained text for most purposes. Supplemental discussion along these lines and on multicomponent systems can be found in such books as "Phase Diagrams in Metallurgy," by F. N. Rhines, McGraw-Hill Book Company, 1956; "Alloy Phase Equilibria," by A. Prince, Elsevier Publishing Company, 1966; and "Ternary Systems," by G. Masing and B. A. Rogers, Reinhold Publishing Corporation, 1944.

I am indebted to Dr. Lester Guttman for clarifying discussions on the order-disorder transition in alloys and for the development of the mathematics in Chap. 5 relative to the quasi-chemical treatment of this transition, and to Professor M. B. Bever for helpful suggestions relative to the treatment of free-energy surfaces of ternary diagrams in Chap. 8. I am also grateful to the many authors and copyright owners for their permissions to use the various illustrations from their publications which I have incorporated in this book. And, finally, I thank Mr. J. R. Dvorak for supplying the photomicrograph in Fig. 4.11, Dr. R. A. Rideout and Dr. T. R. Pritchett for the photomicrograph in Fig. 4.12, Professor C. S. Barrett for the x-ray photograms in Fig. 5.2, Dr. L. F. Mondolfo for the photomicrographs in Figs. 6.12, 6.14, and 6.15, Mr. J. R. Villela and Dr. H. C. Knechtel for the photomicrographs in Fig. 6.22, Mr. T. Watmough for the white-iron sample from which the photomicrographs in Fig. 6.24 were made, and Professor F. N. Rhines for the photomicrograph in Fig. 7.4.

Paul Gordon

CONTENTS

LIST OF SYMBOLS

E	Internal energy
Q	Heat absorbed by system
W	Work done by system
S	Entropy
T	Temperature
P	Pressure
V	Volume
G	Free energy
H	Enthalpy
μ	Chemical potential
X	Mole, or atom, fraction
$\bar{E}$, $\bar{S}$, etc.	Partial molal quantities
α, β, γ, etc.	Greek lowercase letters used to represent phases
β	Also used to represent compressibility
A, B, etc.	Roman capital letters used to represent components
C_P	Heat capacity at constant pressure
c	Number of components
p	Number of phases
ln	Logarithm, base e (Naperian, or natural, logarithms)
log	Logarithm, base 10 (common logarithms)
M	As superscript, means mechanical mixture
L	A liquid phase
S	As superscript, a solid solution phase
m	As superscript, means quantity at the melting point
Z	Coordination number
N	Avogadro's number
R	The gas constant
k	Boltzmann's constant (in a few instances k is used as the distribution coefficient, but is defined in each of these cases)
$\mathcal{U}_{AA}$	Interaction energy between two nearest-neighbor A atoms
$\mathcal{U}_{BB}$	Interaction energy between two nearest-neighbor B atoms
$\mathcal{U}_{AB}$	Interaction energy between two nearest-neighbor A and B atoms
n_{AA}, n_{BB}, n_{AB}	The number of AA, BB, and AB nearest-neighbor pairs, respectively

$\mho$	$\mho_{AB} - \left(\dfrac{\mho_{AA} + \mho_{BB}}{2}\right)$
T_c	Critical temperature
S	Long-range order parameter
σ	Short-range order parameter
ΔS_m	Entropy of mixing (the configurational entropy)
$\Delta S^{xs}, \Delta H^{xs}, \Delta G^{xs}$	Excess entropy (exclusive of ΔS_m), enthalpy, and free energy accompanying formation of a solution
Q-C	Quasi-chemical
B-W	Bragg-Williams

In general, when a quantity is intended to refer to a specific phase or component, the symbol representing the phase appears as a superscript, that representing a component as a subscript; e.g., $\mu_B{}^\alpha$ is the chemical potential of the component B in the phase α.

PRINCIPLES OF PHASE DIAGRAMS
IN MATERIALS SYSTEMS

ONE

THE NATURE AND
IMPORTANCE OF PHASE DIAGRAMS

Materials in the solid state exist in many different forms, or *phases*. The number of such phases can be large even for pure substances; e.g., ice may exist in any one of six different solid forms, and the important metal iron exhibits four solid phases. In systems containing more than one *chemical species* the number of phases may be correspondingly larger; not infrequently even in commercially important systems the available data cover only parts of the systems. In view of the fact that the properties of materials depend significantly on the nature, number, amounts, and forms of the various possible phases present and can be changed by alterations in these quantities, it is vital in the use of materials to know the conditions under which any given system will exist in its various possible forms. A wealth of such information on a large number of systems has been accumulated. To record this enormous amount of data, it has become customary to plot the number and compositions (and, indirectly, the amounts) of phases present as a function of temperature, pressure, and overall composition. These plots are called *phase diagrams*, *constitution diagrams*, or *equilibrium diagrams*. The latter name is derived from the fact that such diagrams purport to show the most stable phases that occur under equilibrium conditions.

The term "equilibrium conditions" requires some elaboration. Strictly speaking, the *equilibrium state* of a system is that state in which the properties of the system will not change with time ad infinitum unless acted upon by some constraint. In this sense constraint usually means an alteration in temperature, pressure, or composition but may also refer to the application of mechanical, electric, etc., forces. In practice, however, the definition of equilibrium is generally modified to take into account the relative rates of the possible processes which may be induced in a system by changes in temperature, or pressure, or composition. Frequently a system is of importance only within a certain range of temperature, pressure, and composition. In such a case reactions within the system may proceed at rates which are (1) so slow that they produce only negligible property changes in the longest practical time periods, (2) so fast that they proceed to equilibrium in less than the shortest practical time periods, and (3) of intermediate magnitude. The phase diagram will usually show the reactants in type (1) processes and the products of reaction in type (2) and (3) processes as the *equilibrium phases*. The most notable example of this is the iron-carbon phase diagram. This diagram includes the important system of alloys known as steel, upon which, it is no exaggeration to say, our modern civilization is founded. The solid phases commonly shown in the iron-carbon diagram are three forms of iron (α-iron, γ-iron, and δ-iron) and cementite, or iron carbide, Fe_3C. The latter, if given enough time, ultimately decomposes to graphite and iron; nevertheless, iron carbide is generally shown in the diagram in preference to graphite because in the range of temperature, pressure, and composition important to steel, iron carbide is the carbon-rich phase almost invariably encountered.

Today, both the determination and the use of equilibrium diagrams are largely empirical. There is, however, a firm basis for such diagrams in the science of thermodynamics. In principle, at least, equilibrium diagrams can be calculated from thermodynamic relationships; in practice, however, this can be done as yet only for some relatively simple situations, for two reasons: the exact theory for the general case is prohibitively complex, and much of the necessary fundamental thermodynamic data are not yet available. It is, nevertheless, quite instructive to consider equilibrium diagrams from a thermodynamic point of view, for such an approach allows a critical insight into many of the qualitative features of diagrams and also provides a basis for understanding the areas into which the science of equilibrium diagrams is now advancing. In this book, therefore, the thermodynamic viewpoint is taken as the basis of discussion, elaborated upon where needed by the more conventional empirical, or geometrical, treatment.

TWO

THERMODYNAMIC FUNDAMENTALS

2.1 STATE QUANTITIES

Thermodynamics is a broad scientific discipline the essence of which is the development of useful quantitative mathematical relationships between the measurable properties of systems. The development of these relationships is based on observational and experimental evidence as to the nature of systems and their reactions. The relationships are expressed in terms of quantities which we may designate as *state variables* and *state properties*.

In an operational sense *state variables* are the potentials which can be manipulated in a reasonably direct fashion to fix or alter the condition, or *state*, of a system, i.e., those quantities through which constraints may be exerted on systems. Most frequently the temperature, pressure, and composition are the only state variables of significance; in special situations others, such as mechanical, electric, magnetic, gravitional, and surface potentials, may be important. For example, in making an alloy in a laboratory furnace, the composition of the alloy and the pressure under which it is heated determine the temperature at which melting occurs; the exact gravitational potential at the particular furnace location as compared with that at some other possible location is unimportant. On the other hand, if it became necessary to determine whether this alloy would melt as a result of frictional and impact heating after being dropped from some given height above the earth's surface, then obviously the gravitational potential with respect to the earth would be of paramount importance. In the treatment of phase diagrams only the temperature, pressure, and composition are normally of importance; hence only these variables will be considered here.

Physical chemistry, wherein thermodynamics had its origins, has traditionally dealt with gases and liquids. Experience has shown that with these substances when the state of a system has been fixed by regulating the temperature, pressure, and composition, all the properties of the system are uniquely specified to a high degree of precision, regardless of the past history of the system.* The properties of such substances are called state properties; since these properties are unique functions of the state, they are also single-valued functions of the state variables temperature, pressure, and composition. They can, therefore, be handled quantitatively in the mathematical expressions of thermodynamics. In dealing with materials, which are almost invariably solids, it is frequently found that the properties depend as much on the past history as on the temperature, pressure, and composition at the time of interest. For example, many of the properties of a steel bar, particularly its mechanical properties, are completely unspecified if only its temperature, pressure, and composition are given. The entire history of the bar from the time the steel was cast into an ingot has a bearing on its properties. This situation is primarily due to two characteristics of solids, their ability to resist shear stresses and the relative sluggishness of reactions within solids as compared with gases and liquids. These characteristics of solids are responsible for many of their useful properties, but in the study of equilibrium diagrams they must be kept in mind as limitations on the application of thermodynamic equations.†

2.2   A CRITERION OF EQUILIBRIUM

Thermodynamics has made it possible to express, in a few short principles, or laws, the evidence of innumerable observations and experiments as to the nature of the universe. These laws, expressed

* This is true only so long as the particle size in the system does not approach atomic dimensions.

† It is probable that the establishment of complete internal equilibrium within solids would virtually eliminate the influence of history on the properties, rendering them state properties, but this concept is not universally accepted.

mathematically, form an extremely useful set of equations relating the state quantities of systems.

The *first law* of thermodynamics asserts that for a system in internal equilibrium there is a quantity, called *internal energy E*, which is a state property of the system. More specifically, this law states that whenever a closed* system is carried from a given initial state to a given final state by any path, or process, whatsoever, if all the heat Q absorbed by the system and all the work W done by the system during the process are measured, the difference between Q and W will be constant for all paths. This is true in spite of the fact that the magnitudes of both Q and W separately *do depend* on the nature of the path. In other words, the difference, $Q - W$, depends only on the initial and final states of the system and is, thus, the change of some state property of the system. Since Q and W are both energy terms, this state property is also an energy term; it is given the name internal energy and the symbol E; thus

$$\Delta E_{1 \to 2} = Q - W \qquad (2.1)$$

Internal energy is not described in detail by classical thermodynamics, since the latter treats any single mass of a given homogeneous system as a continuum, i.e., not in terms of discrete particles, or atoms, within the system. Atomistically, E is made up of the potential and kinetic energies of the particles in the system. It should be noted, however, that neither from a thermodynamic nor an atomistic point of view can E, or any other form of energy, be completely defined. That is to say, the magnitude of any energy term can never be given on an absolute basis but only relative to some arbitrary frame of reference. For example, E is only partially defined thermodynamically by Eq. (2.1) or its differential equivalent

$$dE = \delta Q - \delta W \qquad (2.2)$$

(the differential symbol d here and throughout is considered to apply only to changes in state quantities, or exact differentials, and the symbol δ to nonstate changes, or inexact differentials). Equation (2.2) defines only a *change* in E during a process in terms of the two measurable changes δQ, the heat absorbed by the system, and δW, the work done by the system, during the process. If

---

* A closed system is one which undergoes no mass interchange with its surroundings.

Eq. (2.2) is integrated between limits, there results

$$E_2 - E_1 = \int_1^2 \delta Q - \int_1^2 \delta W$$

That is, the energy E_2 of the system in state 2 can only be given in terms of the energy E_1 in state 1 and the change in energy in going from state 1 to state 2. The same limitation exists in defining E atomistically. The kinetic energies of the atoms, for example, are functions of the atomic velocities; since the velocities can only be defined relatively—relative to the earth, to the sun, etc.—the kinetic energies can only be defined relatively. However, it is only energy changes which are of scientific and practical importance; thus, arbitrary reference energies may be used in writing equations containing energy terms. In calculating energy changes these reference energies cancel out.

The _second law_ of thermodynamics asserts that another quantity associated with all systems, called _entropy S_, is also a state property and has the important characteristics that

_1_ In any process taking place for a system in equilibrium with its surroundings (a reversible process) the entropy of the system plus that of the surroundings is a constant, that is

$$dS_{\text{syst}}^{\text{rev}} + dS_{\text{surr}}^{\text{rev}} = 0 \tag{2.3}$$

_2_ In a spontaneous (irreversible) process

$$dS_{\text{syst}}^{\text{irrev}} + dS_{\text{surr}}^{\text{irrev}} > 0 \tag{2.4}$$

This quantity, entropy, is (partially) defined by thermodynamics as

$$dS = \frac{dQ^{\text{rev}}}{T} \tag{2.5}$$

Equations (2.3) to (2.5) may be summarized in the following statement of the second law:

$$dS \geqq 0 \tag{2.6}$$

for a system and its surroundings, i.e., an _isolated*_ system.

* An isolated system, it may be seen, is a portion of the universe which can be considered to undergo no mass, energy, or work interchange with the rest of the universe, i.e., an adiabatic, isometric, closed system.

Again, thermodynamics does not describe entropy in detail, and an understanding of Eq. (2.5) appears frequently to be a stumbling block for the student. When viewed atomistically, the concept of entropy is really no more abstruse than that of energy. Once it is granted that in a conglomeration of atoms each atom has thermal kinetic energy and an associated velocity which continually change both in magnitude and direction by chance collisions with other atoms, then it should be apparent that from a strictly statistical point of view an essentially random positional arrangement of the atoms has by far the highest probability of occurring at any given moment. This tendency toward randomness is perfectly universal, but it is opposed by, and generally at least partly counterbalanced by, the forces derived from the potential energies of the atoms. Nevertheless, the *tendency* toward randomness is always present and is the stronger, the greater the atomic kinetic energies, i.e., the higher the temperature. The thermodynamic quantity entropy is simply an expression of the degree of this randomness of a system, and the second law of thermodynamics is merely a statement of the universal, spontaneous tendency toward complete randomness, or maximum entropy.

It is clear that the actual equilibrium (most probable) arrangement of any group of atoms is determined primarily by two factors, namely, the strength of the interatomic potential energies and the strength of the tendency toward randomness. Those atoms with the highest mutual attractions will tend to be closest together, producing the greatest possible negative interaction energies.* On the other hand, the kinetic energies of the atoms will still make for complete randomness in atom positions, or the highest positive entropy. The optimum compromise between these two opposing tendencies will be the equilibrium state. A quantitative thermodynamic equivalent of this statement may be derived in the following way: combining Eqs. (2.2) and (2.5), we have for a reversible process, i.e., for a system at equilibrium,

$$dE = T \, dS - \delta W \tag{2.7}$$

If, for the time being, we restrict the system under consideration to one of fixed composition and mass for which the only work done

---

* Attractive energies must be defined as negative to agree with the implicit assumption made in stating the first law, namely, that heat *absorbed by* the system and work *done by* the system are taken as positive.

is against pressure, δW becomes $P_{surr}\, dV$, and since $P_{surr} = P_{syst}$ in a reversible process, then

$$dE = T\, dS - P_{surr}\, dV = T\, dS - P\, dV \tag{2.8}$$

The thermodynamic statement in Eq. (2.8) is not in its most useful form, since the change in the dependent quantity E is expressed in terms of the two independent variables S and V, both of which are difficult to manipulate experimentally. This difficulty may be eliminated by the introduction of a new quantity G, called the Gibbs *free energy*, and defined as

$$G \equiv E + PV - TS* \tag{2.9}$$

Differentiation of Eq. (2.9) and substitution into (2.8) yields

$$dG = V\, dP - S\, dT \tag{2.10}$$

Here, the dependent quantity, G, is expressed in terms of two independent variables, P and T, which are readily controlled in the laboratory. For a process carried out at *constant temperature and pressure*, Eq. (2.10) tells us that under equilibrium conditions

$$dG = 0 \tag{2.11}$$

In addition, it may be shown from Eqs. (2.4) and (2.9) that for a spontaneous, or irreversible, process at constant temperature and pressure

$$dG < 0 \tag{2.12}$$

(see Ref. 2.1).† Thus, the first two laws of thermodynamics reveal that in a spontaneous process at constant temperature and pressure, the free energy decreases and reaches a minimum at equilibrium. Atomistically, the tendency for strongly attracting atoms to cluster together lowers both E and V, and the tendency toward randomness raises S. It may be seen from the definitional equation for G, Eq. (2.9), that both of these spontaneous tendencies lower G, in agreement with (2.12).

* The symbol $\equiv$ is used for equalities which are wholly definitional.
† References are listed at the end of each chapter.

2.3   PHASES AND COMPONENTS

The criteria for equilibrium expressed in Eqs. (2.8), (2.10), and (2.11) have been derived for a *closed* system of *fixed composition* for which the only energy changes with the surroundings are due to heat transfer and to work against pressure. In the study of materials, however, we are more frequently interested in systems which may also undergo compositional changes by interchange of matter internally. To facilitate the treatment of such systems, some discussion of the terms *phase* and *component* is desirable.

A specific system may at equilibrium be homogeneous or heterogeneous. In the former case the system is uniform in properties* throughout and contains only one physical boundary, that between itself and its surroundings. Such a system is called a single-phase system. A heterogeneous system contains two or more homogeneous systems, or phases, each separated from the others by distinct physical boundaries.

Every system is made up of a number of chemical species, and it should be noted that, from a thermodynamic point of view, every phase in a system at equilibrium contains some (no matter how small an amount) of each chemical species. (This will be demonstrated later.) The amounts of the chemical species are related to each other through chemical interactions; if, in a given system, these interactions are assumed to be at equilibrium, then the minimum number of chemical species which may be independently varied in amount and by means of which the compositions of every phase in the system may be completely specified are called *components*. In many materials systems the number of components is simply equal to the number of elements in the system. In complex systems, however, the number of components may not be immediately apparent and in practice will depend on the particular experimental conditions of interest, as does the definition of equilibrium previously discussed. For a detailed discussion of components the student is referred, for example, to Refs. 2.1 and 2.2; two illustrative examples will be given here. Consider first a system composed of the chemical species H_2, O_2, and H_2O. At ordinary pressures and temperatures, where H_2O may exist as a liquid, there are two phases in the system, one a liquid consisting of an aqueous solution of oxygen and hydrogen, the other a vapor containing the two gases and water vapor.

---

* Uniform on a microscopic and macroscopic scale. No system is completely uniform on an atomic scale, but thermodynamics considers systems to be made up of continua on any scale.

Under these conditions any chemical reaction between H_2, O_2, and H_2O is extremely slow. For most purposes, then, the effect of this reaction on the state of the system is negligible. Thus at low temperatures all three species, O_2, H_2, and H_2O, are independently variable, and all must be specified to fix the compositions of the two phases. The system is then properly considered to consist of three components. At high temperatures, however, where the reaction

$$2H_2 + O_2 \rightleftharpoons 2H_2O$$

proceeds to equilibrium rapidly, the system H_2, O_2, and H_2O contains only two components. When the quantities of any two of the three species present are specified, that of the third is no longer independently variable but is fixed by nature, i.e., by the equilibrium relationship for the above reaction.

As a second example consider the system iron carbide, Fe_3C, in equilibrium with a solid solution of carbon in solid iron, e.g., in steel. The Fe_3C is here not strictly a stable phase, as was pointed out earlier. At high enough temperatures or after long enough times Fe_3C decomposes to iron and graphite. Under most conditions of interest in steel, however, this reaction proceeds slowly enough to be neglected, so that Fe_3C may be considered to be a stable species. The reaction between Fe_3C and carbon in solution in the solid iron

$$Fe_3C \rightleftharpoons Fe + C$$

does, however, proceed essentially to equilibrium at temperatures where steels are normally heat-treated. Thus, the amounts of only two of the three chemical species involved may be independently varied, and hence only two may be considered to be components. The choice of which two are selected is quite arbitrary, though for convenience the elemental species Fe and C are usually chosen.

2.4 EQUILIBRIUM IN SYSTEMS OF VARIABLE COMPOSITION

We are here interested in multicomponent, multiphase, i.e., heterogeneous, systems which, though still of fixed overall composition and mass, may undergo internal composition variations by the transfer of matter from phase to phase. Such a system may best be treated by focusing attention first on any single phase within the system. We may change our viewpoint for the moment and look upon this phase as the system under consideration, the other phases being part of the surroundings. To be consistent with the condition that in the original heterogeneous system matter transport could take place between phases, the new single-phase system must be considered an open system, i.e., one which can interchange matter with its surroundings. Each such matter interchange leads to an internal energy transfer, which, for an infinitesimal change, may be expressed as

$$dE = \sum \frac{\partial E}{\partial n_i} dn_i = \sum \mu_i \, dn_i \tag{2.13}$$

where μ_i, called the chemical potential of i, is the rate of change of the internal energy with change in amount of the ith component, all other variables being held constant.

In writing the first law for an open system of this type the energy terms in Eq. (2.13) must be added to the heat and work terms. As a result, the combined statement of the first and second laws for a reversible process in a single-phase, open system in which heat effects, pressure work, and matter interchange with the surroundings are taken into account becomes

$$dE = T \, dS - P \, dV + \Sigma \mu_i \, dn_i \tag{2.14}$$

In this expression, the chemical potential of each component is taken at constant S, V, and amounts of the other components. Combining expression (2.14) with the differential of the free-energy function, Eq. (2.9), we obtain the alternate expression of the combined first and second laws

$$dG = V \, dP - S \, dT + \Sigma \mu_i \, dn_i \tag{2.15}$$

The chemical potentials in this expression are still written at constant S, V, and amounts of the other components. Since, however, dG is an exact differential, it can also be written

$$dG = \frac{\partial G}{\partial P} dP + \frac{\partial G}{\partial T} dT + \sum \frac{\partial G}{\partial n_i} dn_i \tag{2.16}$$

where the partial of G with respect to a given variable is in each case now written with the other variables of the group T, P, n_1 n_2, . . . held constant. Comparison of (2.13), (2.15), and (2.16) reveals that

$$\left(\frac{\partial G}{\partial n_i}\right)_{T,P,n_1...n_{i-1},n_{i+1},...} = \mu_i = \left(\frac{\partial E}{\partial n_i}\right)_{S,V,n_1...n_{i-1},n_{i+1},...} \tag{2.17}$$

In view of Eq. (2.17) the chemical potentials in (2.15) can refer to the partial of the free energy at constant T, P, n_1, n_2, . . . , as well as the partial of the internal energy at constant S, V, n_1, n_2, . . . , and we may write simply

$$dG = V\,dP - S\,dT + \mu_1\,dn_1 + \mu_2\,dn_2 + \cdots \tag{2.18}$$

without reference to constancy subscripts.

In deriving expressions (2.14) and (2.18) for a single phase of the multiphase system originally under consideration no restriction was applied as to which particular phase in the system was being discussed. Thus, an expression of the type (2.18) can be written for each phase in a multiphase system. For the system as a whole the analogous expressions (2.8) and (2.10) apply, since the overall heterogeneous system is of constant mass and composition.

The treatment of multiphase systems at equilibrium does not usually invoke any difficulty in acceptance of the concept that all the phases must have the same temperature and pressure. It is, perhaps, less apparent but equally true that each component is characterized by a chemical potential that is the same in all phases. This may be demonstrated as follows. Consider any two phases, α and β, within a system at equilibrium. If an amount dn_1 is transferred from phase α to phase β holding T, P, and the amounts of the other components constant, then from (2.18)

$$dG^\alpha = \mu_1{}^\alpha\,dn_1$$

and

$$dG^\beta = -\mu_1{}^\beta\,dn_1$$

The total change in free energy of the two phases is

$$dG = dG^\alpha + dG^\beta = (\mu_1{}^\alpha - \mu_1{}^\beta)\,dn_1$$

Since, however, the overall composition and mass of the two phases together have not changed, these two phases can be viewed as a system of fixed composition and mass which is in equilibrium at constant temperature and pressure. Thus, Eq. (2.11) applies, and

$$dG = (\mu_1^\alpha - \mu_1^\beta)\, dn_1 = 0$$

or

$$\mu_1^\alpha = \mu_1^\beta \tag{2.19}$$

A similar argument can be advanced with regard to any other two phases in the system and for every component. As a consequence, it is seen that there is only one chemical potential, characteristic of the system as a whole, for each component in a system at equilibrium.

Expressions (2.14) and (2.18) are forms of the combined statement of the first and second laws of thermodynamics for a reversible (equilibrium) process in a homogeneous system which may interchange matter with its surroundings. Another equivalent form, which we shall find useful, may be derived in the following way. The internal energy of a phase at fixed values of the temperature, pressure, and composition may be found by integrating Eq. (2.14)

$$dE = T\, dS - P\, dV + \Sigma \mu_i\, dn_i \tag{2.14}$$

Since the temperature and pressure are fixed, and the chemical potentials, which at constant T and P are functions only of the composition, are also constant, then

$$E = TS - PV + \Sigma \mu_i n_i + \text{const} \tag{2.20}$$

Complete differentiation of (2.20) gives

$$dE = T\, dS + S\, dT - P\, dV - V\, dP + \Sigma \mu_i\, dn_i + \Sigma n_i\, d\mu_i \tag{2.21}$$

Upon substraction of (2.14) from (2.21) we obtain

$$0 = S\, dT - V\, dP + \Sigma n_i\, d\mu_i \tag{2.22}$$

which is the required expression.

Problems

2.1 Show by consideration of a schematic P-V diagram for a process in which the only work done is against pressure that the total work done is necessarily a function of the path by which a system is taken from some initial state A to some final state B.

2.2 The first law of thermodynamics is sometimes stated as follows. "When a system is taken by any process from some initial state A through any other state B and back to its initial state A, the internal energy change is zero." From this statement alone, show that the energy change in going from state A to state B is independent of the path.

2.3 If a system proceeds spontaneously from a state 1 to a state 2, what relationship does the total heat effect accompanying this process have to the entropy change, $\Delta S = S_2 - S_1$, of the system? Why?

2.4 Under what conditions does each of the following criteria of equilibrium apply: (a) $\Delta G = 0$; (b) $\Delta S = 0$?

2.5 Consider a system consisting of coke (carbon) in an atmosphere of CO plus CO_2. For most purposes, how many components are present (a) at room temperature and (b) at temperatures where the coke will react readily with the gases?

2.6 Describe a constant-temperature, constant-pressure process in which a system is taken from some initial state to some different final state with no change in free energy. Can such a process be carried out in the laboratory?

2.7 By consideration of the dynamic atomic interchanges at the interface between two phases in a two-component system, show that in general in such a system two contiguous phases at equilibrium must have different compositions.

2.8 What is the relationship between the chemical potential of a component in a phase and the quantities involved in the equilibrium in Prob. 2.7?

References

2.1 L. S. Darken and R. W. Gurry, "Physical Chemistry of Metals," McGraw-Hill Book Company, New York, 1953.

2.2 A. Findlay, A. M. Campbell, and N. O. Smith, "The Phase Rule," p. 9, Dover Publications, Inc., New York, 1951.

THREE

ONE-COMPONENT SYSTEMS

3.1 TEMPERATURE–PRESSURE–FREE-ENERGY DIAGRAM IN ONE-COMPONENT SYSTEM

One-component systems are by definition pure substances. Such systems, though still of relatively limited importance in the study of materials, have been receiving greater attention in recent years, particularly with the enhanced interest in high-pressure reactions. Because of this trend and because the simplicity of one-component systems makes them especially suitable for the study of principles applicable to all systems, one-component systems will be treated in some detail. In view of the free-energy criteria stated in expressions (2.10) to (2.12), the study of equilibrium relationships in one-component systems may be logically started by a consideration of the temperature–pressure–free-energy diagram.

For each phase in any closed system of fixed overall composition an equation similar in form to Eq. (2.20) may be written for the free energy, namely,

$$G = f(T, P, X_1, X_2, \ldots) \tag{3.1}$$

since G is a state property and is thus a function only of temperature, pressure, and composition. In (3.1), G is taken as a molal quantity, so that the amount of each component is represented by its mole fraction X. (Subsequently in this book all such quantities will be considered to be molal quantities unless otherwise stated.) When this has been done, minimization of the total free energy of the system with respect to phase compositions at each temperature, pressure, and overall composition gives the phase or group of phases stable under each condition, i.e., the equilibrium phase diagram. For a one-component system, Eq. (3.1) is simplified to

$$G = f(P,T) \tag{3.2}$$

and, at equilibrium, this function can be found from Eq. (2.10)

$$dG = V \, dP - S \, dT \tag{2.10}$$

Integration of (2.10) from some (arbitrary) reference point $G(P_0, T_0)$ on the surface $G = f(P,T)$ gives

$$G(P_1, T_1) = G(P_0, T_0) + \int_{P_0}^{P_1} V(P, T_0) \, dP - \int_{T_0}^{T_1} S(P_1, T) \, dT \tag{3.3}$$

In this equation V, the molal volume, can be measured as a function of pressure, so that the first integral can be evaluated graphically. The second integral can also be evaluated graphically, from experimental data on the variation with T of the heat capacity at constant pressure C_P. The latter is defined as

$$C_P \equiv \left(\frac{\partial Q}{\partial T} \right)_P \tag{3.4}$$

Combining Eqs. (3.4) and (2.5), we see that at constant P

$$dS = \frac{C_P}{T} \, dT \tag{3.5}$$

so that

$$\int S \, dT = \iint \frac{C_P}{T} \, dT \, dT$$

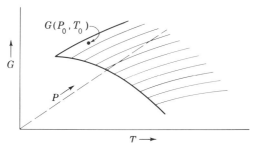

Fig. 3.1   *Schematic surface in G-T-P space.*

Thus, at least in principle, the function G in Eq. (3.3) can be calculated. Graphically, this produces a surface in G-T-P space, as shown schematically in Fig. 3.1, with the values of G being relative to the arbitrary reference point $G(P_0, T_0)$.

In a one-component system there is a free-energy surface such as this for each phase—the vapor phase, the liquid phase, and one or more solid phases. These surfaces intersect each other, and that portion of each surface which is lowest of all surfaces in any region gives the T-P range in which the corresponding phase is stable. A diagram of this kind showing the lowest free-energy portion of the G-T-P surfaces is depicted schematically in Fig. 3.2 for the simplest of one-component systems, in which there is assumed to be only one possible solid phase. The intersection lines of the free-energy surfaces are one-dimensional regions of two-phase stability; i.e., each such line is a region where two phases have equal free energies and thus can coexist in equilibrium. Each point of intersection of three lowest free-energy surfaces produces a zero-dimensional region of three-phase stability called a *triple point*. When these lines and points are projected into a T-P plane, the pressure-temperature equilibrium phase diagram is produced. In Fig. 3.2 the lines CD, DE, and PD project into the basal plane to become the phase diagram given by the lines $C'D'$, $D'E'$, and $P'D'$.

3.2   RELATIONSHIP BETWEEN FREE-ENERGY SURFACES AND PROPERTIES OF THE PHASES

The surfaces in Fig. 3.2 are quite simply related to many of the properties of the corresponding phases. From Eq. (3.2) differenti-

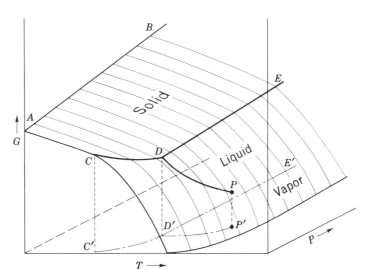

Fig. 3.2   *Schematic G-T-P diagram for a one-component system, showing the portions of the solid, liquid,
and vapor free-energy surfaces which represent lowest
free-energy conditions.*

ation gives

$$dG = \left(\frac{\partial G}{\partial P}\right)_T dP + \left(\frac{\partial G}{\partial T}\right)_P dT \tag{3.6}$$

and comparing this with Eq. (2.10),

$$dG = V\, dP - S\, dT \tag{2.10}$$

it is seen that

$$\left(\frac{\partial G}{\partial P}\right)_T = V \tag{3.7}$$

and

$$\left(\frac{\partial G}{\partial T}\right)_P = -S \tag{3.8}$$

Thus, the slopes of the free-energy surfaces in isothermal planes through Fig. 3.2 are equal to the molal volumes of the respective phases. This means that these slopes are all positive in the P direction, and, except at very high pressures, those of the solid and liquid surfaces are quite small compared to that of the vapor surface. In the case of aluminum, for example, the molal volume of the solid at ordinary room temperature and pressure is about 10 cm³, and that of the liquid a few percent greater, whereas that of the vapor would be 24 liters. Similarly, the slopes of the free-energy surfaces in isobaric planes are equal to the entropies of the respective phases. Since entropies are positive at finite temperatures, and in general

$$S_{vap} > S_{liq} > S_{sol}$$

these slopes are also positive and have the above relationship in magnitude. The orders of magnitude in this case are approximately the same, however, for all three phases. Taking aluminum at its melting point and atmospheric pressure as an example, the entropies of the solid, liquid, and vapor are about 14, 17, and 41 cal/(°K) (mole), respectively.

The curvatures of the free-energy surfaces in Fig. 3.2 are directly related to the compressibilities and the heat capacities of the phases. Thus, differentiating Eq. (3.7) at constant temperature,

$$\left(\frac{\partial^2 G}{\partial P^2}\right)_T = \left(\frac{\partial V}{\partial P}\right)_T = -\beta V \qquad (3.9)$$

where the compressibility β is defined by

$$\beta \equiv -\frac{1}{V}\left(\frac{\partial V}{\partial P}\right)_T \qquad (3.10)$$

Since both β and V are positive, these constant-temperature curvatures are negative, and all three surfaces in Fig. 3.2 are concave downward in isothermal planes. Again, however, since solids and liquids are highly incompressible as compared with vapors, the solid and liquid surfaces in Fig. 3.2 are almost flat in the P direction, whereas that for the vapor is highly curved. In a like manner, from Eq. (3.8)

$$\left(\frac{\partial^2 G}{\partial T^2}\right)_P = -\left(\frac{\partial S}{\partial T}\right)_P \qquad (3.11)$$

and combining this with Eqs. (2.5) and (3.4), it may be seen that

$$\left(\frac{\partial^2 G}{\partial T^2}\right)_P = -\frac{C_P}{T} \tag{3.12}$$

These isobaric curvatures are also negative, since C_P and T are both positive, and hence the free-energy surfaces are concave downward in the constant-pressure, as well as the constant-temperature, direction. The magnitudes of the curvatures are comparable for solid, liquid, and vapor in this case, since the heat capacities of the three states of aggregation are not too different.

The relationships developed above are summarized in Table 3.1.

Table 3.1   **Relationship between Free-energy Surfaces and Some Properties of the Phases in One-component Systems**

Free-energy surface characteristic	Equivalent property of phase	Sign	Relative magnitude		
			Solid	Liquid	Vapor
SLOPE, $\left(\dfrac{\partial G}{\partial P}\right)_T$	V	positive	small	small	large
SLOPE, $\left(\dfrac{\partial G}{\partial T}\right)_P$	$-S$	negative	of the same order		
CURVATURE, $\left(\dfrac{\partial^2 G}{\partial P^2}\right)_T$	$-\beta V$	negative	small	small	large
CURVATURE, $\left(\dfrac{\partial^2 G}{\partial T^2}\right)_P$	$-\dfrac{C_P}{T}$	negative	of the same order		

3.3   EQUILIBRIUM PHASE DIAGRAM FOR SIMPLE ONE-COMPONENT SYSTEM. THE PHASE RULE

The projected lines $C'D'$, $D'E'$, and $D'P'$ in Fig. 3.2 are redrawn in the more familiar T-P phase diagram of Fig. 3.3.* This diagram consists of:

1   Three areas within each of which the system at equilibrium exists as a single phase, either solid, liquid or vapor, as the case may be

* Although in many physical chemistry books, P, rather than T, is taken as the vertical ordinate in T-P diagrams, the opposite convention is adopted in this book to be consistent with common usage in multicomponent, constant-pressure diagrams.

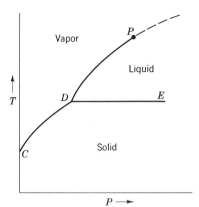

Fig. 3.3   *The   T-P   equilibrium   phase   diagram
derived from G-T-P diagram in Fig. 3.2.*

2   Three lines along each of which two phases coexist at equilibrium:
the line *CD* for solid-vapor, *PD* for liquid-vapor, and *DE* for solid-liquid

3   A point *D* at which all three phases, solid, liquid, and vapor,
coexist at equilibrium

It may be noted that there is a systematic (and inverse) relationship
between the dimensionality of a given *T-P* region and the number
of phases which may coexist at equilibrium in that region. The
dimensionality of a region is, in a one-component system, equal to
the *number of degrees of freedom*, designated *f*, of the region, i.e., the
number of variables which for a system at equilibrium may be arbitrarily specified (within limits) without the disppearance of a phase
or the appearance of a new phase. Although the identity between *f*
and the dimensionality of a phase region carries over to multicomponent systems only for single-phase regions, the relationship
between *f* and the number of phases coexisting at equilibrium is
quite general; because of its generality and usefulness it has been
given a special name, the *phase rule*. Written in terms of the number
of components in a system *c* and the number of phases *p* coexisting
at equilibrium in a given phase region, the phase rule states that

$$f = c - p + 2 \tag{3.13}$$

Resort to such a rule may seem hardly necessary in a system as
simple as that in Fig. 3.3, where only two possible variables, tem-

perature and pressure, are involved. In systems containing two or more components, however, phase diagrams become considerably more complex, as will become apparent later; the phase rule is then an important guiding rule. It will, therefore, be derived here, and examples of its use will be given throughout the subsequent discussion.

The phase rule results essentially from an application to thermodynamic equilibrium of an elementary mathematical principle which may be stated as follows: If a group of n variables are related by m independent conditions, the number of arbitrarily alterable variables, i.e., the number of degrees of freedom f, is

$$f = n - m$$

Every student has made use of this principle, perhaps implicitly, innumerable times, e.g., in his acceptance of the fact that m algebraic equations are required to fix uniquely the values of n algebraic variables. To derive the phase rule, it is necessary simply to express the number of thermodynamic variables n in a general system at equilibrium and the number of conditions of equilibrium m in terms of the number of components c in the system and the number of phases p coexisting under any given set of equilibrium conditions, and then to subtract the one from the other. The variables of concern here are the temperature, the pressure, and the composition variables as expressed through the chemical potentials. Since in any system at equilibrium the temperature, pressure, and chemical potential of each component are uniform throughout the system, the total number of variables n is $c + 2$ In each phase these variables are related at equilibrium by the combined first and second laws of thermodynamics, as given in Eq. (2.22). There are p phases and therefore p such restrictions, or equations, connecting the variables; that is, $m = p$. Thus, the number of degrees of freedom is

$$f = n - m$$
$$f = c - p + 2$$

Returning to the schematic diagram in Fig. 3.3, it may now be seen that, in the terminology usually associated with the phase rule, the one-phase regions in this one-component diagram may be described as *divariant*, the two-phase regions as *univariant*, and the three-phase region as *invariant*. This means that if the system concerned is stipulated to exist at a temperature and pressure such that it consists at equilibrium of

1 A single phase, then both the temperature and pressure at which this single phase will exist may be arbitrarily selected within the limits of the appropriate field in the diagram;

2 Two phases, then only one variable, the temperature or the pressure, may be arbitrarily selected (again within limits), the other being fixed by nature along the appropriate two-phase line in the diagram;

3 Three phases, then neither the temperature nor the pressure may be arbitrarily selected: both are fixed by nature at the point D in the diagram. Furthermore, only one such point is possible for any three phases.

The nature of the three two-phase lines in Fig. 3.3 may be viewed in several equivalent but usefully different ways. As mentioned above, each represents the combination of temperature and pressure at which two phases may coexist in equilibrium; it may also be said that each such line indicates the temperature-pressure combinations at which a given phase becomes saturated with respect to another or the temperature-pressure combinations at which the transformation of one phase to another can take place at equilibrium. The transformations have been given the names *sublimation* and *condensation* for that from solid to vapor and its reverse, respectively, along the line CD, *vaporization* and *condensation* along the line PD, and *melting* and *solidification* along DE. The sublimation line CD also gives the equilibrium vapor pressure of the solid as a function of temperature, and the line PD that for the liquid.

The observation that three phases in equilibrium in a one-component system produce an invariant situation at a triple point can be generalized by reference to the phase rule. Thus, in any system, an invariant situation exists when f is zero, so that the number of phases to bring about this condition is

$$p = c + 2$$

This is also the maximum number of phases which may coexist at equilibrium in any system, since $f = 0$ is the smallest value of f possible, negative values being physically meaningless.

The triple point, D in Fig. 3.3, is of further interest in the sense that its position with respect to 1 atm pressure determines whether

a substance normally sublimes or melts on heating from the solid
state. For most substances, D is at a pressure well below 1 atm;
heating at atmospheric pressure produces melting in these substances. For a few substances, however, D is *above* atmospheric
pressure; e.g., it is at 5.1 atm (and $-56.6°C$) in the case of CO_2.
The latter, therefore, sublimes on normal heating, giving the
phenomenon of "dry ice."

3.4   THE LE CHATELIER PRINCIPLE AND THE
CLAUSIUS-CLAPEYRON EQUATION

As may be inferred from the discussion of Fig. 3.2, the slopes of the
equilibrium transition lines in phase diagrams are closely related,
both in sign and magnitude, to the *relative* properties of the phases
concerned. A qualitative expression of this relationship is given in
the *Le Chatelier principle* and a quantitative statement in the
Clausius-Clapeyron equation.

The Le Chatelier principle, which is universally applicable
and extremely useful in varied problems, states that "when a constraint is exerted on a system in equilibrium, the system changes in
such a way as to tend to relieve the constraint." Accordingly, since
the liquid and solid phases of a given substance are generally denser
than their equilibrium vapors, the lines CD and PD in Fig. 3.3 must
always have positive slopes. This may be visualized physically by
imagining, say, the vapor and solid in equilibrium at a temperature
and pressure corresponding to some point on CD and asking what
would happen if the pressure were raised to a somewhat higher value
at constant temperature. In order to relieve the constraint, the
higher pressure, the system would tend to lower its volume; it
can do this by transforming vapor, the high-volume phase, to
solid, as long as vapor remains. When all the vapor is gone, the
system will consist of a single phase, solid, and will thus be represented in the diagram of Fig. 3.3 by a point lying within the solid-
phase field. Since this point was reached by moving from the line
CD to higher pressure at constant temperature, the line CD of
necessity has a positive slope.

The melting-point line DE in Fig. 3.3 can have, on the other
hand, either a positive or a negative slope, depending on whether
the solid or the liquid has a greater density at any given point along
DE. For example, in the H_2O system, shown in Fig. 3.4 at low

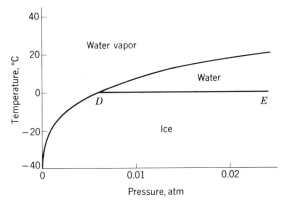

Fig. 3.4 *The H_2O equilibrium phase diagram at low pressures. (Data from "Handbook of Chemistry and Physics," Chemical Rubber Publishing Co.)*

pressures, the melting line *DE* has a (small) negative slope since liquid water in equilibrium with ordinary ice is denser than the ice. This is an anomalous situation, however, most materials being denser in the solid than in the liquid state. It seems worth noting in passing that this anomalous behavior of water can be considered as a very significant accident of nature, for it is unlikely that without it life as we know it would exist. The earliest forms of life presumably took shape in the relatively friendly environment of large bodies of water. If ice were denser than water, however, this primitive life would undoubtedly have been snuffed out by settling ice during the first big freeze, rather than finding a haven in the unfrozen water remaining under the floating ice fields. In a less serious vein, the pleasure of ice skating might also never have been experienced. Ice is actually not very slippery; however, when a skater supports his weight on the sharp blade of a skate, a very large pressure is developed just under the blade edge. This *lowers* the melting point of the ice locally, the resulting water forms a thin film of lubricant between the blade and the ice, and the almost frictionless glide essential for ice skating becomes possible.

It may be noted that the slopes of the transition lines *CD* and *DP* in Fig. 3.3 are quite high; i.e., the sublimation temperature and the boiling temperature are highly dependent on pressure. The line *DE*, on the other hand, has a very small slope; the melting temperature is relatively insensitive to pressure. This is a consequence of the fact that the PV term in Eq. (2.9)

$$G \equiv E + PV - TS \tag{2.9}$$

is negligible for condensed phases unless the pressure is very high but is quite significant for gases even at low pressures. In other words, the lowest free-energy (equilibrium) state for condensed phases is determined largely by the temperature at low pressures, whereas with gases the pressure on the system is also important. A demonstration of the validity of this statement may be made by means of Eq. (2.10)

$$dG = V\, dP - S\, dT \tag{2.10}$$

The relative effects of pressure and temperature changes on the free energy of a substance may be calculated from the ratio of the two terms $V\, dP$ and $S\, dT$. Applying the equation to water at 0°C and 1 atm pressure and assuming, say, a 10% change in both pressure and temperature, we find

$$\frac{S\, dT}{V\, dP} = \frac{(0.9)(0.1 \times 273)(41.3)}{(1)(0.1 \times 1)} \cong 12{,}000$$

In making this calculation, the specific volume of water was taken as 1 cm³/g, the entropy of water at 0°C and 1 atm was found from tables of thermodynamic data to be 0.9 cal/(°C)(g), the quantity 41.3 appears as the conversion factor from calories to cm³-atm, and it has been assumed that over the small range of T and P concerned both S and V of water may be approximated as constant. The result clearly indicates that at ordinary temperatures and pressures changes in the free energy of liquid water due to pressure changes are negligible compared to the effect of temperature. If, on the other hand, gaseous H_2O were being considered, the specific volume would be larger by a factor of about 10^3, and the two terms $S\, dT$ and $V\, dP$ would then be more nearly equal.

The magnitudes of the transition-line slopes may be calculated by means of the *Clausius-Clapeyron equation*. This equation results from a combination of Eq. (2.10) with the fact that along each transition line the free energies of the two phases of concern are equal. Thus

$$G^\alpha = G^\beta$$

and

$$dG^\alpha = V^\alpha \, dP - S^\alpha \, dT = dG^\beta = V^\beta \, dP - S^\beta \, dT$$

$$\frac{dT}{dP} = \frac{\Delta V}{\Delta S} \tag{3.14}$$

which is a form of the Clausius-Clapeyron equation. An alternate form results if use is made of the quantity H, called enthalpy, and defined as

$$H \equiv E + PV \tag{3.15}$$

Combining this with Eq. (2.9) gives

$$G = H - TS \tag{3.16}$$

and since the free energies of the two phases are equal at the transition temperature, then at this temperature

$$\Delta G = \Delta H - T \, \Delta S = 0$$

or

$$\Delta S = \frac{\Delta H}{T}$$

Substituting this into (3.14) gives

$$\frac{dT}{dP} = \frac{T \, \Delta V}{\Delta H} \tag{3.17}$$

In this form of the Clausius-Clapeyron equation, dT/dP is the slope of the transition line of concern, ΔV is the difference in the volumes of the two phases, and ΔH^* is the heat effect of the transition, all at a specific temperature and pressure. As an example of the use of this equation, the slope of the melting-point line DE at atmospheric pressure in the H_2O phase diagram (see Fig. 3.14) may easily be calculated since in this case ΔH is 79.7 cal/g, the specific volume of ice is 1.090 cm³/g, and that of water is 1.000 cm³/g; therefore

$$\frac{dT}{dP} = \frac{(273.3)(-0.090)}{(79.7)(41.3)} = -0.0075°C/atm$$

---

* ΔH, the difference in enthalpy, is, at constant pressure, equal to the heat effect Q, as may be seen from Eqs. (2.2) and (3.15).

Table 3.2   **Experimental and Calculated Initial (Atmospheric-Pressure) Slopes of Melting-temperature–Pressure Curves of Metals[3.1]**

Metal	Struc-ture*	Volume,† cm³/ mole	ΔS,† cal/ (mole)(°K)	$\frac{\Delta V†}{V}$	dT/dP, °K/kbar — Calcu-lated	Experimental
LITHIUM	(bcc)	13.3	1.59	0.0165	3.3	3.3, 1.5
SODIUM	(bcc)	24.1	1.68	0.025	8.6	7.8, 6.5
POTASSIUM	(bcc)	46.0	1.65	0.0255	16.9	13.3, 8.5
RUBIDIUM	(bcc)	56.1	1.79	0.025	18.7	18.0
CESIUM	(bcc)	70.0	1.69	0.026	25.7	20.0, 19.0
ALUMINUM	(fcc)	10.5	2.74	0.060	5.5	6.4
COPPER	(fcc)	7.6	2.30	0.0415	3.3	4.2
SILVER	(fcc)	10.9	2.19	0.038	4.5	5.5
GOLD	(fcc)	10.7	2.21	0.051	5.9	
NICKEL	(fcc)	7.1	2.44	0.037	2.6	3.7
PLATINUM	(fcc)	9.5	2.30	0.038	3.8	5.0
RHODIUM	(fcc)	8.7	2.32	0.039	3.5	5.9
LEAD	(fcc)	18.9	1.90	0.035	8.3	10.0, 6.6
IRON	(fcc)	7.7	2.20	0.032	2.7	3.0
IRON	(bcc)	7.7	2.03	0.030	2.7	
THALLIUM	(bcc)	17.8	1.77	0.022	5.3	9.0, 5.0
THALLIUM	(fcc)	17.8	1.98	0.023	4.9	
MAGNESIUM	(hcp)	14.8	2.31	0.041	6.3	7.5
ZINC	(hcp)	9.5	2.55	0.042	3.7	4.5, 4.8
CADMIUM	(hcp)	13.4	2.44	0.040	5.3	9.0, 5.6
INDIUM	(tet)	16.2	1.82	0.020	4.3	4.8, 5.6
TIN	(tet)	16.5	3.41	0.028	3.2	2.8, 4.3, 2.7
TELLURIUM	(hex)	21.0	5.78	0.020	1.7	0.1
ANTIMONY	(rh)	18.7	5.25	−0.0095	−0.8	−0.50, −0.44
BISMUTH	(rh)	21.5	4.78	−0.0335	−3.6	−3.8
GERMANIUM	(dc)	13.9	6.28	−0.050	−2.7	−3.3
GALLIUM	(or)	11.8	4.41	−0.032	−2.0	−2.1

* bcc = body-centered cubic; fcc = face-centered cubic; hcp = hexagonal close-packed; dc = diamond cubic; hex = hexagonal; tet = tetragonal; rh = rhombohedral; or = orthorhombic.
† $\Delta S = S^L - S^S$; $\Delta V = V^L - V^S$; $V = V^S$.

Thus, it requires an increase in pressure of well over 100 atm to lower the melting point of ice by 1°C!

The initial (atmospheric-pressure) slopes of the melting-temperature–pressure lines have been measured for a number of metals. The values are listed in Table 3.2 along with values calculated from Eq. (3.14) and data for the molar volumes, volume changes, and entropy changes at the atmospheric melting points. The agreement between calculated and measured initial slopes is, with a few exceptions, quite satisfactory. It is interesting to note that antimony, bismuth, germanium, and gallium are anomalous in the same sense as water; i.e., the liquids are denser than the solids. As a result the melting-line slopes are negative. These four elements are in many respects only semimetals. They lie in Groups III, IV, and V in the periodic table and, typically, have *crystal bonding forces* which are partially metallic and partially nonmetallic. As a result their *crystal structures* are less symmetrical and less close-packed than those of the normal metals; breaking down these structures by melting actually allows the atoms to be, on the average, closer together, producing the negative ΔV's shown in Table 3.2.

3.5   ALLOTROPY, OR POLYMORPHISM.
HIGH-PRESSURE EFFECTS

The fundamental difference between liquids and solids is that the former consist of essentially random arrays of particles (atoms or molecules) whereas the latter are made up of particles arranged in regular patterns; i.e., solids are crystalline. Since with indistinguishable particles only one type of spatially random array is possible but crystalline arrays may be of many types, in a pure material only one liquid phase can exist, but there may be a number of different solid phases. The various crystalline forms in a given material are called *allotropes*, or *polymorphs*, and the transition from one form to another *allotropic*, or *polymorphic, transformation.*

Allotropy can come about either as a result of change in temperature or change in pressure. Some substances exhibit allotropy on heating at ordinary pressures; iron is a typical example, as shown in Fig. 3.5. When heated at atmospheric pressure, iron transforms at 910°C from the α-iron form (body-centered cubic, bcc) to the γ-iron form (face-centered cubic, fcc), and at 1392°C it transforms again to δ-iron (bcc). The γ-α transformation on cool-

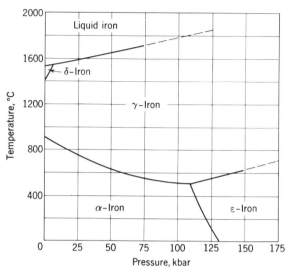

Fig. 3.5   *The iron T-P equilibrium phase diagram.*
(From Takahashi and Bassett, The Composition of the
Earth's Interior. Copyright © 1965 by Scientific
American, Inc. All rights reserved.)

ing is without doubt the most important allotropic transformation
known, since it is responsible for the ability of Fe–C alloys (steel)
to attain their unusually high strengths, without which our modern
civilization could not have been developed.

The thermodynamic requirements for the appearace of allotropy on changing temperature at constant pressure may be deduced
from Eq. (3.8)

$$\left(\frac{\partial G}{\partial T}\right)_P = -S \tag{3.8}$$

Any two possible solid phases, α and β, in a given substance will in
general have different free energies and entropies at a specific
temperature and pressure. The stable phase at low temperatures,
say α, is that with the lower free energy at these temperatures. If,
however, α also has the lower entropy, then as the temperature is
raised, the free energy of β will, according to Eq. (3.8), fall more
rapidly than that of α. This is shown in Fig. 3.6, the diagrams of
which are essentially isobaric sections through the solid-liquid
regions of *G-T-P* plots such as those in Fig. 3.2. Above some tem-

perature T_{tr} in Fig. 3.6, G^β is less than G^α, and β becomes the stable phase if, as indicated in Fig. 3.6a, T_{tr} is below the melting point T_m. In this case β will be stable in the temperature range $T_{tr} < T < T_m$, and the substance exhibits allotropy. Whenever the free-energy relationships are as in Fig. 3.6b, melting will occur before β can become stable, and allotropy does not appear on increasing the temperature.

Although changes in pressure as compared with changes in temperature have only relatively small effects on the free energy of solids, with the advent of very high-pressure techniques it has been found that many solids undergo phase transitions at constant temperature when the pressure is increased to sufficiently high values.

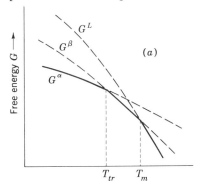

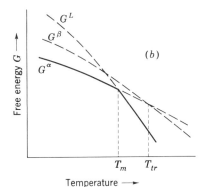

Fig. 3.6   *Schematic free-energy–temperature curves (constant pressure) illustrating the conditions leading to temperature-induced allotropy. The portions of the curves corresponding to phase stability are drawn in full lines; the solid phases are indicated by α and β, the liquid by L. Allotropy appears in (a) but not in (b).*

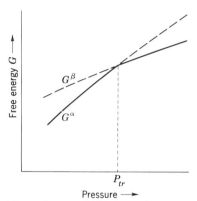

Fig. 3.7 *Schematic free-energy–pressure curves (constant temperature) illustrating the conditions leading to pressure-induced allotropy. Phase stability indicated by full lines.*

The pressures needed are of the order of tens or hundreds of *kilobars* (1 kbar = 10^6 dynes/cm² $\cong$ 1000 atm). Both the possible occurrence of allotropy on increasing pressure and the order of magnitude of the pressures needed are predictable from Eqs. (2.10) and (3.7). From Eq. (3.7) it is seen that the free energy of a phase is increased as the pressure is increased at constant temperature, and the rate of increase is equal to the volume V of the phase. Since two solid phases, α and β, of the same substance will at a given temperature and pressure have different V's the rate of free-energy change of that phase with the higher V will be the greater. If this phase, say α, is stable at low pressures, it will have the lower G at these pressures, but as the pressure is increased, G^β will ultimately become lower than G^α, as shown in Fig. 3.7, and the substance will exhibit allotropy with pressure increase. The pressure at which the free-energy curves cross is the transition pressure P_{tr}; above this pressure β is the stable (lower-free-energy) phase. The manner in which such curves lead to T-P phase diagrams is illustrated in Fig. 3.8.

The range of pressures necessary to bring about such phase transitions may be estimated if it is assumed that the free-energy change required is roughly the same as that required for temperature-induced allotropy. From Eq. (2.10), this is equivalent to assuming that

$$\int_{P_0}^{P_{tr}} V \, dP \cong \int_{T_0}^{T_{tr}} S \, dT$$

where P_0 and T_0, for example, may be taken as atmospheric conditions. Since only an order-of-magnitude calculation is desired and both V and S change only slowly with P and T, respectively, the integration of this equation may be approximated as

$$P_{\text{tr}} - P_0 \cong \frac{S_{\text{av}}}{V_{\text{av}}} (T_{\text{tr}} - T_0)$$

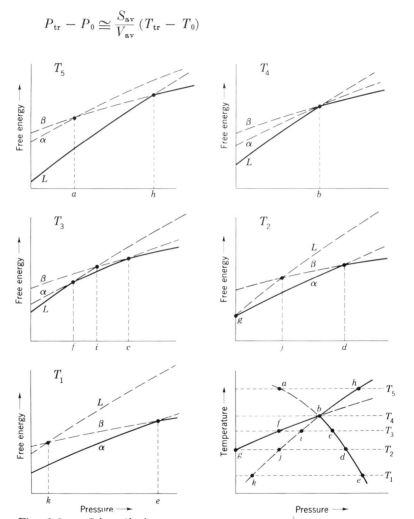

Fig. 3.8 *Schematic free-energy–pressure curves at several temperatures and the resulting T-P phase diagram in a hypothetical system for which $V^L > V^\alpha > V^\beta$ and $S^L > S^\beta > S^\alpha$. Full lines correspond to equilibrium phases and phase transitions.*

where S_{av} is the average entropy over the temperature interval $T_{tr} - T_0$ and V_{av} is the average volume over the pressure interval $P_{tr} - P_0$. The ratio S_{av}/V_{av} is not too different for many metals, and $T_{tr} - T_0$ is typically of the order 10^2 to 10^3°K. Taking iron as an example, $T_{tr} - T_0$ for the α-γ transition is about 900°K, and S_{av}/V_{av} is about 70 atm/°K, so that

$$P_{tr} - P_0 \cong 70 \times 900 \cong 6 \times 10^4 \text{ atm} \cong P_{tr}$$

This means that for iron, and many other metals, if pressure-induced allotropy is to be found, the transition pressures should be expected to be of the order 10 to 100 kbar, as is generally observed.

Viewing the atoms in solids as incompressible spheres, the above discussion suggests that pressure-induced allotropy in solids is probable whenever the solid phase stable at atmospheric pressure is not ideally close-packed. Increased pressure then tends to favor formation of a denser phase, i.e., one with the atoms filling space more efficiently; at pressures on the order of 10 to 100 kbar, the denser phase may become stable. This, in fact, has recently been found to be the case for iron, as shown in Fig. 3.5. The α form of iron stable at room temperature and pressure is bcc, not quite close-packed. On increasing the pressure to 130 kbar at room temperature, a new phase ε, which is hexagonal close-packed (hcp), is observed to form.

In most metals the atomic structures have virtually ideal close packing; on the basis of a hard-sphere model, therefore, high-pressure transformations in most metals would appear unlikely. In some metals, such as α-iron, the structure is not close-packed; in these, and particularly in the semimetals, which have quite open structures (see discussion page 29), high-pressure transformations are more probable. Bismuth is typical of the semimetals, and, correspondingly, phase transitions at high pressures have in fact been found for bismuth. This is shown in the phase diagram reproduced in Fig. 3.9. High-pressure transformations have also been observed in, for example, barium, calcium, strontium, antimony, thallium, cerium, and ytterbium.

The hard-sphere view of atoms, useful though it is in many cases, is only a crude approximation; there is little doubt that atoms are compressible to some degree. The high-pressure transformation in ytterbium is an illustration of this, in that it has been found to produce a high-pressure bcc structure from a low-pressure fcc

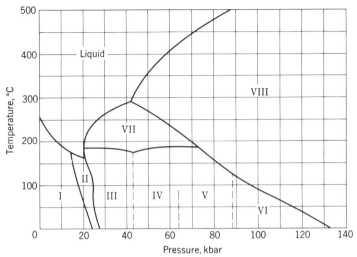

Fig. 3.9   *Equilibrium phase diagram for bismuth.*
(*Adapted from Strong.*[3.3])

structure.[3.4] In the hard-sphere view of atomic packing this is not possible, for on this basis a bcc structure should be somewhat less dense than an fcc structure in the same material. The apparent discrepancy has been rationalized[3.5] by a theory which takes into account the finite compressibility of atoms. In this theory it is proposed that at high pressures the compressive forces acting between two neighboring atoms may be expected for a given pressure to be inversely proportional to the coordination number. Since the coordination number, the number of equidistant neighbors, is 12 for fcc packing (and also for hexagonal close-packing) and 8 for bcc packing, there will be some pressure where the normally more voluminous bcc form will become denser than the fcc form; a transformation from the fcc to the bcc structure results. It is suggested[3.5] that such a situation may be expected in other elements as well as in ytterbium. If this theory is correct, it would mean that not only may high-pressure phases form from close-packed low-pressure phases, but in some instances it would be possible for increasing pressure to change a non-close-packed phase to a close-packed phase and then subsequently back to a non-close-packed phase.

Allotropy can be induced not only by increasing pressure or

temperature but also by combinations of both high temperatures
and high pressures. An indication of the combinations presently
attainable is given in Fig. 3.10. The application of such temperature-
pressure combinations appears to present the most interesting and
potentially useful possibilities. This is due in part to the fact that
phases so produced can sometimes be retained at atmospheric pressure and temperature. One of the first examples of such a new phase
was the production of a new form of boron nitride, called Borazon.
Boron nitride in its natural form has properties quite similar to
graphite, but Borazon is so hard that it can actually scratch diamond, the only material other than diamond itself which is known
to have this ability. Furthermore it is considerably more oxidation-
resistant than diamond, and thus its potential usefulness is quite
high.

High temperature and pressure combinations have also been
able to produce diamond artificially from graphite. The *T-P* range
where diamond exists stably is shown in the carbon phase diagram
of Fig. 3.11. It is apparent that at ordinary temperatures and pressures diamond is unstable, though the time necessary for its reversion
to graphite under these conditions is sufficiently long to reduce to

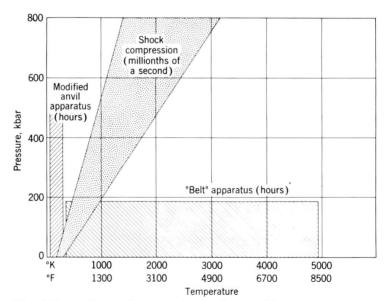

Fig. 3.10   *Range of pressure-temperature combina-*
tions attainable (1964). (From Suits.[3.6]*)*

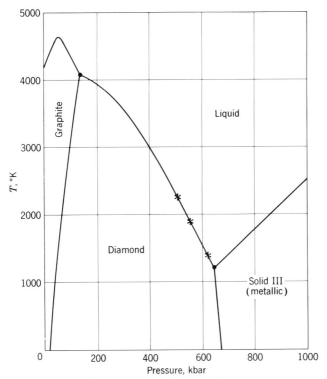

Fig. 3.11   *The carbon equilibrium T-P diagram.*
(*From Suits.*[3.6])

zero the danger that it will do so in any except geologic time scales.
It may also be seen from Fig. 3.11 that heating graphite at atmospheric pressure (in a nonoxidizing atmosphere) simply produces
melting, but if the heating is done at sufficiently high pressures,
diamond is formed. When the diamond is then cooled to room
temperature under pressure and the pressure ultimately released,
the artificially produced diamond remains in the metastable condition just like natural diamond.

The effects of high pressure on the transformations of condensed
phases have been quantitatively calculated in some one-component
systems on the basis of Eq. (3.3) and data for the variation with
temperature and pressure of the volumes and entropies of the two
respective phases involved. Results of two such calculations for iron
and tin and comparison with experiment are given in Figs. 3.12
and 3.13.

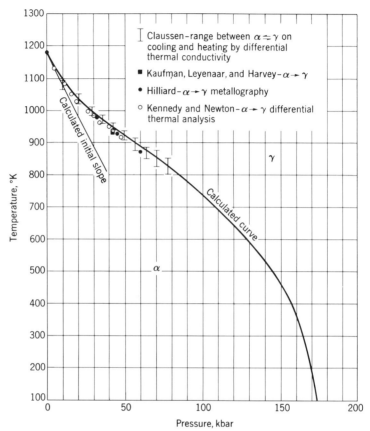

Fig. 3.12   *Effect of high pressure on the α-γ transformation in iron. (From Kaufman.[3.1])*

3.6   THE DEGREE OF A TRANSFORMATION

Phase transformations such as those across lines *CD*, *DE*, and *DP* in Fig. 3.3 are accompanied by discontinuous changes in many properties; e.g., the densities of the phases involved are different at equilibrium and, especially, there are *latent heats* of transformation. Transitions of this kind are designated *first-order*, or *first-degree*, transitions. A different kind of transition is indicated by the fact that the line *DP* in Fig. 3.3 ends within the diagram, at a point *P*, called the *critical point*. At temperatures and pressures beyond this point the vapor-liquid transition is no longer discontinuous; i.e., there is no boiling point and no latent heat. Attempts have been made to classify these and other transitions by defining the degree of the transition thermodynamically in terms of the derivatives of the free energy with respect to temperature (at constant pressure) for the phases involved in the transformations; in particular, the degree of the transformation is said to correspond to the lowest degree of the free-energy derivatives for which there is a finite difference at the equilibrium temperature for the two phases concerned. According to this system of classification, thus, a first-degree transformation is one for which

$$(\Delta G)_{T_E, P} = 0 \tag{3.18}$$

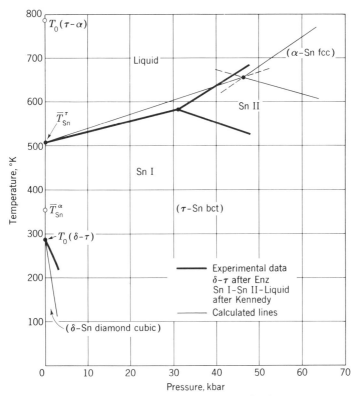

Fig. 3.13   *Experimental and calculated equilibrium T-P diagrams for tin. (From Kaufman.[3.1])*

but

$$\Delta \left(\frac{\partial G}{\partial T} \right)_{T_E,P} \neq 0 \tag{3.19}$$

where T_E indicates the equilibrium transformation temperature at the pressure P. A second-degree transition is one for which (3.18) holds but

$$\Delta \left(\frac{\partial G}{\partial T} \right)_{T_E,P} = 0 \tag{3.20}$$

and

$$\Delta \left(\frac{\partial^2 G}{\partial T^2} \right)_{T_E,P} \neq 0 \tag{3.21}$$

and a third-degree transition is characterized by (3.18), (3.20), and the two additional equations

$$\Delta \left(\frac{\partial^2 G}{\partial T^2} \right)_{T_E,P} = 0 \tag{3.22}$$

and

$$\Delta \left(\frac{\partial^3 G}{\partial T^3} \right)_{T_E,P} \neq 0 \tag{3.23}$$

and so on.

This classification has in practice not proved particularly fruitful, however, because it has generally been found that transformations of higher than first degree do not have all the characteristics implied by equations of the type (3.20), (3.21), etc. It may well be more useful to limit the thermodynamic classification by designating a transformation only as either first degree or higher degree simply on the basis of whether or not it is characterized by a latent heat. By the use of Eqs. (3.8), (3.18), (3.19), and (3.20) we may easily show that for a first-degree transition

$$(\Delta H)_{T_E} \neq 0 \tag{3.24}$$

whereas for a second- or higher-degree transformation

$$(\Delta H)_{T_E} = 0 \qquad\qquad\qquad (3.25)$$

Following this categorization, the transformations across lines CD, DE, and DP in Fig. 3.3, are of first degree but the vapor-liquid change beyond the critical point is of higher degree.

From a physical point of view there is, perhaps, an even more striking difference between first-degree and non-first-degree transformations. In a first-degree transition both phases will exist together in equilibrium once the transformation is under way; e.g., in boiling along line DP in Fig. 3.3, the vapor and liquid phases will be found to coexist with a distinct physical boundary between them, and if the addition of the heat producing the boiling is halted, they will continue to coexist ad infinitum. In a non-first-degree transition, on the other hand, there is at equilibrium no point at which two phases coexist; there is simply a single phase present at all times and as the temperature (or pressure) is changed, the properties of the phase gradually, continuously, and uniformly change from those of the original material to those of the final material. In Fig. 3.3, as the temperature is raised at a constant pressure beyond the critical point, the liquid gradually adopts the properties of the vapor, but at no time do the two phases exist together in equilibrium. In this sense, only transformations of first degree should be called phase transformations, those of higher degree being more properly looked upon as changes of the properties of a single phase.

The critical point in the transition between liquid and vapor is typically at pressures of tens or hundreds of atmospheres. This is illustrated in the phase diagram for H_2O in Fig. 3.14, where the critical point has the coordinates 374°C, 218 atm. Solid-liquid transition lines, however, extend to pressures as high as have been experimentally obtainable (of the order of hundreds of kilobars, see Fig. 3.10). That is to say, there is always a discontinuous change between solid and liquid. There seems to be no conclusive theoretical proof either for or against the existence of a critical point between solid and liquid. However, from an atomistic viewpoint, it seems apparent that such a critical point should be at least considerably less probable than that between liquid and vapor. As pointed out, the atomic arrangements in liquid and vapor are both essentially random, the transformation between the two involving, therefore, mainly a change in density. Intuitively it is expected that the difference in density may disappear at high pressures. Since solids are crystalline with very regular atomic arrays, however, the transition between solid and liquid involves a distinct rearrangement in structure; such a rearrangement might be expected to give rise to a discontinuous transition even at high pressures.

Transformations of high degree in substances other than water are of greater interest in materials science. For example, both the ferromagnetic transformation in α-iron and certain order-disorder transformations in alloys are transitions of this type. The specific-heat–versus–temperature and enthalpy–versus–temperature curves for iron at constant (atmospheric) pressure are shown in Fig. 3.15. It may be seen that the α-γ, γ-δ, and δ-liquid transformations are all first-degree transitions accompanied by finite changes in H at constant temperature and pressure. The ferromagnetic transformation, however, takes place over a range of temperatures, most rapidly near 760°C; it produces no discontinuous change in H at any temperature but rather an unusually rapid change in C_P, particularly near 760°C; it is thus not a phase (first-degree) transformation. It

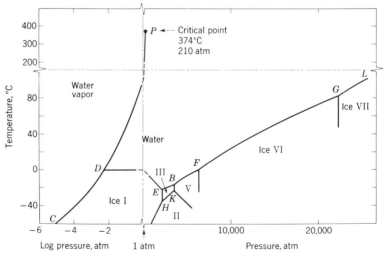

Fig. 3.14   *The H_2O equilibrium phase diagram. (Low-pressure data from "Handbook of Chemistry and Physics," Chemical Rubber Publishing Co.; high-pressure diagram adapted from Findlay, Campbell, and Smith.*[2.2]*)*

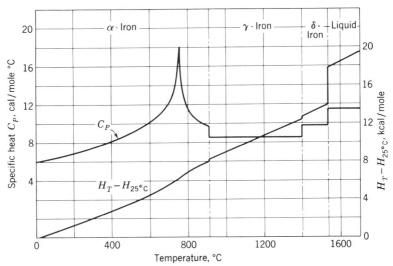

Fig. 3.15   *Molal specific heat and enthalpy of iron at atmospheric pressure. (Data from ASM "Metals Handbook.")*

is of historical interest to note that the existence of the ferromagnetic transformation is responsible for the absence of the term β-iron in the sequence α, γ, and δ used for designating the solid phases of iron. When the heat effect resulting from the rapid disappearance of ferromagnetism over the temperature range near 760°C was first observed, it was thought to be due to a phase transformation, the new phase being designated β-iron. This designation was simply dropped when the true nature of the transition was later delineated.

The order-disorder transition is treated in Chap. 5. As will be seen, it may have the characteristics either of a first-degree or a higher-degree transition and in some cases the characteristics of both.

Problems

3.1   The specific heat of solid copper above room temperature is given by

$$C_P = 5.41 + 1.50 \times 10^{-3}T \qquad \text{cal/mole}$$

If the entropy of solid copper at 300°K is taken as 8.00 cal/(mole) (°K) what is the entropy at 1073°C?

3.2   Show by the Le Chatelier principle that (*a*) the transformation from a low-temperature phase to a high-temperature phase at constant pressure must always be accompanied by a positive heat effect (heat absorbed by system); (*b*) the transformation from a low-pressure phase to a high-pressure phase at constant temperature must always be accompanied by an increase in density.

3.3   A block of copper is contained in a gastight chamber at room temperature and at 1 atm pressure of some inert gas. The chamber is then heated under essentially equilibrium conditions to 1000°K. If it is desired to know the free-energy change of the copper to an accuracy of $\pm 1\%$, show that the change in free energy due to the pressure change in the chamber is negligible. Make any reasonable assumptions and approximations in arriving at your answer.

3.4   Assuming that the differences in volume ΔV and enthalpy ΔH between liquid and solid of a pure substance are not functions of pressure or temperature, calculate and plot the melting-point–versus–pressure curves for iron up to 100 kbar and for bismuth up to 20 kbar. Compare with the experimental curves in Figs. 3.5 and 3.9. Comment on the agreement between calculated and experimental curves.

3.5 γ-Iron is fcc, and ε-iron is hcp. Suggest a method by which this knowledge and the iron *T-P* diagram could be used to estimate the stacking-fault energy in γ-iron.

3.6   Draw schematic *G-P* curves which would be consistent with the iron *T-P* diagram in Fig. 3.5 at the temperatures 1600, 910, 600, 500, and 300°C (500°C is the triple-point temperature between α, γ, and ε). Neglect the δ phase.

3.7   It has been suggested that the compressibilities of isomorphs at high pressures may be inversely proportional to the coordination number Z. If, accordingly, it is assumed as a first approximation that, for α-iron and γ-iron, the compressibility relationship is $\beta^\alpha = 1.5\beta^\gamma$ (since $Z^\alpha = 8$ and $Z^\gamma = 12$), that the compressibilities are independent of temperature and pressure, and that at atmospheric pressure $V^\alpha = 1.02V^\gamma$ (independent of temperature), calculate the pressure above which α-iron should become denser than γ-iron. What would be the slope dT/dP of the α-γ transformation line at this pressure? Compare with Fig. 3.5. (The compressibility $\beta^\alpha = 5.9 \times 10^{-7}/\text{atm}$.)

3.8   Derive Eqs. (3.24) and (3.25).

3.9   Derive the relationships for the heat capacity at T_E analogous to Eqs. (3.24) and (3.25) for second- and third-degree transitions.

3.10   During the equilibrium solidification of liquid bismuth at 1 atm pressure the extraction of heat is stopped when 50% of the metal is solidified. If the temperature is now held constant and the pressure increased, what will happen? On what basis is this prediction made? How much of an increase in pressure is required to complete whatever process is produced by the increase?

3.11 Answer the questions in Prob. 3.10 for the solidification of liquid iron.

3.12 If the solid and liquid phases of a pure substance have the enthalpies and entropies listed below when they are in equilibrium with each other, what is the melting temperature?

$$H_T^{sol} - H_{25°C}^{sol} = 8000 \text{ cal/mole}$$
$$H_T^{liq} - H_{25°C}^{sol} = 10,000 \text{ cal/mole}$$
$$S^{sol} = 8 \text{ cal/(mole)(°K)}$$
$$S^{liq} = 12 \text{ cal/(mole)(°K)}$$

3.13 Richard's rule (an empirical rule) states that the entropy of fusion is approximately a constant from substance to substance. Plot the ΔS data in Table 3.2 versus melting temperatures. From this plot, what can you say as to the general validity of Richard's rule? Does the plot suggest a relationship between the validity of Richard's rule and the periodic table? If Richard's rule is assumed to be valid, what relationship exists between melting temperature and heat of fusion?

3.14 Suppose you have a system consisting of the chemical elements A, B, C, and D and you know that under the conditions of temperature, pressure, and time for which you will be using the system the following two reactions go substantially to equilibrium:

$$2A + B \rightleftharpoons A_2B$$
$$3C + 5D \rightleftharpoons CD_2 + C_2D_3$$

No other reactions take place between the various chemical species. (a) How many components are there in this system under these conditions? (b) How many degrees of freedom would there be in a two-phase region in this system? (c) What would be the maximum number of phases which would coexist at equilibrium in this system?

References

3.1   L. Kaufman, Some Equilibria and Transformations in Metals under Pressure, in Paul and Warschauer (eds.), "Solids under Pressure," McGraw-Hill Book Company, New York, 1963.

3.2   T. Takahashi and W. A. Bassett, *Sci. Am.*, **212**(6):100 (1965).

3.3   H. M. Strong, *Am. Scientist*, **48**:58 (1960).

3.4   H. T. Hall, J. D. Barnett, and L. Merrill, *Science*, **139**:111 (1963).

3.5   F. K. Weddeling, *J. Appl. Phys.*, **35**:1201 (1964).

3.6   C. G. Suits, *Am. Scientist*, **52**:395 (1964).

FOUR

TWO-COMPONENT, ISOMORPHOUS SYSTEMS

4.1 ISOBARIC SECTIONS

In two-component systems there are three thermodynamic variables, the temperature, the pressure, and one composition variable. Thus the complete equilibrium diagram is a three-dimensional figure. When, in the simplest case, both components have diagrams of the type shown in Fig. 3.3, are isomorphous, i.e., of the same crystal structure, and mutually soluble in all proportions in the vapor, liquid, and solid states, the two-component equilibrium diagram at low pressures looks schematically as shown in Fig. 4.1. All regions in the one-component diagrams add a dimension in the two-component diagram. Points become lines; e.g., points a and b become the line ab; lines become surfaces; e.g., lines ae and bf become the surface $abfe$; and areas become volumes; e.g., the areas gae (the solid region for component A) and hbf (the solid region for component B) become the volume under the two surfaces $abhg$ and $abfe$. Consequently, all regions add one degree of freedom; e.g., the triple point a in the one-component system has $f = 0$, as shown before, but the line ab, which represents the triple-point position as affected by composition, has $f = 1$, since in this case

$$f = c - p + 2 = 2 - 3 + 2 = 1$$

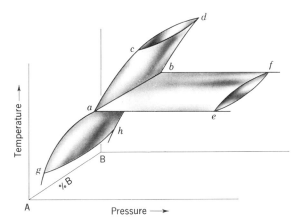

Fig. 4.1   *Schematic temperature-pressure-composition diagram for a simple two-component system at low pressures.*

The treatment of two-component diagrams in three dimensions, particularly when the diagrams become more complex than that in Fig. 4.1, is quite laborious and inconvenient. As a result, it has become customary to treat two-component phase diagrams holding one variable constant, most frequently the pressure; the diagrams then become isobaric sections through constructions such as that in Fig. 4.1. In most materials systems interest lies largely or entirely in the condensed phases at ordinary pressures; in such cases the isobaric sections do not include the vapor-solid and vapor-liquid reactions. Furthermore, since changes in pressure have little effect on the equilibria of condensed phases at ordinary pressures, one isobaric section at or near atmospheric pressure most frequently suffices. For the thermodynamic treatment of two-component systems outlined in this book, we shall for the most part adopt the convenient procedure of considering only the condensed phases at atmospheric pressure; the effect of pressure on the two-component equilibrium diagram will, however, be treated briefly.

4.2   THE FREE ENERGY OF A MECHANICAL MIXTURE AND OF A SOLUTION

Let us examine first a simple binary system in which we stipulate that the only condensed phases possible are the pure components, *A* and *B*, and solutions of the components in each other. Let us further stipulate that each component has only one crystalline form and that the two have the same crystal structure, so that complete intersolubility is possible in the solid, as well as in the liquid, form.* Such a system may exist, at any given temperature, pressure, and overall composition, in any one of the following states:

1 A *mechanical mixture* of the pure components, mechanical mixture being defined as an intimate association of the two pure components in a state of subdivision large enough so that the percent of atoms within the interphase boundaries is negligibly small†

2 A single solution, a solution being defined as a mixture of the components on an atomic scale

3 A mechanical mixture of solutions of different compositions

The problem of delineating the equilibrium diagram thermodynamically consists of predicting which of the above possible states is characterized by the lowest free energy at each temperature, pressure, and overall composition. To facilitate the subsequent discussion we shall adopt the word *alloy* to signify a specific overall composition in the system. In addition, to distinguish a molal property of a component in its pure form from the corresponding property per mole of the component when it is in solution, the *partial molal property*, we shall use a capital letter, say Y, for the former and the same letter barred, $\bar{Y}$, for the latter. It should be noted that in general Y will in fact not be the same as $\bar{Y}$. This is because the environment of, say, A atoms in pure A is not the same as that of A atoms in a solution of B and A. Thus, for example, the internal energy E_A is different from $\bar{E}_A$; this is primarily because the atomic interaction energies around an A atom entirely surrounded by A atoms are different from those around an A atom in the midst of both A and B atoms in solution.

Similarly, the entropy S_A differs from $\bar{S}_A$ because the vibrational frequencies and, thus, the positional randomness of A atoms in

---

* Solid solutions may be substitutional or interstitial. In the former the different atomic species *substitute* for each other *on* the various lattice sites. In the latter the solute atoms reside *in the interstices* between the solvent atoms. Substitutional solutions are by far the more numerous of the two; only in substitutional solutions is complete intersolubility possible.

† In metals and many other materials, intraphase and interphase boundaries are believed to have effective thicknesses of only about one or two atomic dimensions; for most purposes, therefore, the percent of atoms in the boundaries becomes negligible when the particle size of a mixture exceeds 100 to 200 atomic dimensions. It should be noted that this size is well below the best resolution possible with ordinary light microscopes.

pure A are not the same as those of A atoms in solution. In the case of the entropy there is still another important change, called the configurational *entropy of mixing*, or simply the entropy of mixing, attendant on forming a solution. This change is purely statistical in nature; it is due to the fact that there are two kinds of atoms mixed together and is exactly analogous to the change in entropy accompanying the mixing of two gases. Before mixing, the atoms or the gas molecules are in a condition of relatively high order, since each kind of atom is restricted to a specific portion of the system; after mixing, the atoms are in a much more random arrangement, since both kinds of atoms can move throughout the whole system, and the extra randomness is equivalent to an increase in entropy, the entropy of mixing. Since this entropy increment is not associated with either component alone or with individual atoms but rather with the solution as a whole, we shall not include it as part of $\bar{S}_A$ or $\bar{S}_B$ but list it separately as ΔS_m.

With this introduction, we may write for the free energy of a mechanical mixture in any binary system

$$G^M = X_A H_A + X_B H_B - T(X_A S_A + X_B S_B) \tag{4.1}$$

where X_A and X_B are the mole fractons of A and B in the alloy, and H_A, H_B, S_A, and S_B are the enthalpies and entropies of the pure components. For the same alloy existing as a single solution the free energy is

$$G^S = X_A \bar{H}_A + X_B \bar{H}_B - T(X_A \bar{S}_A + X_B \bar{S}_B) - T \Delta S_m \tag{4.2}$$

The enthalpies and entropies in these equations all vary with temperature, pressure, and composition. Since, however, we are dealing only with condensed phases, pressure changes have relatively little effect and will be neglected for the time being, the pressure being assumed fixed unless specifically stated otherwise. We may then examine G^M and G^S as a function of composition at specific temperatures to ascertain the lowest-free-energy conditions.

4.3   THE FREE ENERGY OF DILUTE SOLUTIONS

Consider first the implication of Eqs. (4.1) and (4.2) at either end of the composition range, say for dilute solutions of B in A. If the B atoms are distributed completely at random in the solution, as will be the case in sufficiently dilute solutions, the entropy of mixing per mole is given by

$$\Delta S_m = -R(X_A \ln X_A + X_B \ln X_B) \tag{4.3}$$

(see Sec. 5.4 and Prob. 5.5), and G^S becomes

$$G^S = X_A \bar{H}_A + X_B \bar{H}_B - T(X_A \bar{S}_A + X_B \bar{S}_B)$$
$$+ RT(X_A \ln X_A + X_B \ln X_B) \tag{4.4}$$

In dilute solution the partial molal quantities $\bar{H}_A$, $\bar{H}_B$, $\bar{S}_A$, and $\bar{S}_B$ are all, to a good approximation, not a function of composition, since the environment of individual atoms does not change appreciably on changing the composition. Thus, differentiating G^S with respect to X_B at constant temperature and noting that $X_A = 1 - X_B$,

$$\frac{dG^S}{dX_B} = (\bar{H}_B - \bar{H}_A) - T(\bar{S}_B - \bar{S}_A) + RT \ln \frac{X_B}{1 - X_B} \tag{4.5}$$

As $X_B \to 0$, the quantity $\ln [X_B/(1 - X_B)] \to - \infty$; since $\bar{H}_B$, $\bar{H}_A$, $\bar{S}_B$, and $\bar{S}_A$ are all finite, then at any finite temperature dG^S/dX_B is always negative for sufficiently small values of X_B. This means that the first minute addition of *any* solute to *any* pure substance always lowers the free energy; i.e., under *equilibrium conditions* absolutely pure substances are thermodynamically impossible. Every substance is soluble to some extent in every other substance! It is, therefore, thermodynamically incorrect to draw schematic equilibrium diagrams showing components with a zero range of solubility. In practice, of course, the solubility may be so small that it cannot be shown on the chosen scale of a diagram; thermodynamically it is nevertheless finite.

4.4   THE FREE ENERGY OF IDEAL SOLUTIONS

For compositions other than those adjacent to the pure components, the enthalpies and entropies in Eq. (4.2) all may vary appreciably with composition and temperature. Quantitative information on these properties can in general be obtained only by resort to experiment. We may, however, deduce some of the qualitative and semi-quantitative features of the relevant equilibrium relationships by reference to the useful concept of an *ideal solution* and to some typical deviations from ideality.

An ideal solution is, by definition, one for which

$$\bar{H}_A - H_A = 0 = \bar{H}_B - H_B$$
$$\bar{S}_A - S_A = 0 = \bar{S}_B - S_B \tag{4.6}$$

This implies that the atoms or molecules of the two components are virtually identical chemically and physically but are nevertheless "tagged" in such a way that they can be distinguished from each other. Real solutions are, therefore, never ideal, but in some instances closely approach ideality. The concept is useful in the same sense as is that of an ideal gas. Combining Eqs. (4.1), (4.2), and (4.6), it is seen that the free energy of an ideal solution is

$$G^{id,S} = G^M - T \, \Delta S_m \tag{4.7}$$

Since, in an ideal solution, the interaction energies between the atoms are not dependent on the type of atoms involved, the two kinds of atoms will be distributed in a statistically random fashion, and (4.7) becomes

$$G^{id,S} = G^M + RT[(1 - X_B) \ln (1 - X_B) + X_B \ln X_B] \tag{4.8}$$

The quantity in the bracket is always negative, so that at constant temperature

$$G^{id,S} < G^M$$

for all compositions and all finite temperatures. Thus, at all finite temperatures and over the entire composition range the system exists more stably as a single solution than as a mechanical mixture.

It may also be shown that the single solution is more stable than any mechanical mixture of solutions. The nature of the free-energy–composition curve for an ideal solution at some general temperature T is shown in Fig. 4.2 along with the free energy of the mixture and entropy-of-mixing curves, of which it is the sum.*

---

* In the strictest sense condition (4.6) implies that $H_A = H_B$, $S_A = S_B$, $\Delta H_A^{melting} = \Delta H_B^{melting}$, etc. In Fig. 4.2 and subsequently we are in reality discussing virtually ideal solutions, for which it is assumed that H_A is not quite equal to H_B, $\Delta H_A^{melting}$ is not quite equal to $\Delta H_B^{melting}$, etc., even though condition (4.6) is stipulated to hold.

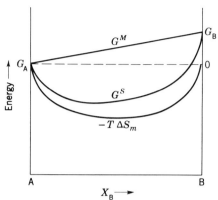

Fig. 4.2   *Free energy of a mixture and the entropy of mixing and free energy of an ideal solution. The free energy of pure A, G_A, is here and in subsequent schematic figures arbitrarily taken to be zero, and that of pure B, G_B, is taken to be greater than G_A.*

The free-energy curve alone is redrawn in Fig. 4.3. Consider now an alloy of overall composition X^1 which is imagined to exist as a mixture of two solutions, one with the composition X^α and the other X^β (see Fig. 4.3). From Eq. (2.18) we may write for the α phase

$$dG^\alpha = V^\alpha \, dP - S^\alpha \, dT + \mu_A{}^\alpha \, dX_A{}^\alpha + \mu_B{}^\alpha \, dX_B{}^\alpha \qquad (4.9)$$

which at constant temperature and pressure becomes

$$dG^\alpha = \mu_A{}^\alpha \, dX_A{}^\alpha + \mu_B{}^\alpha \, dX_B{}^\alpha \qquad (4.10)$$

Noting that $dX_A{}^\alpha = -dX_B{}^\alpha$ and rearranging,

$$\left(\frac{dG}{dX_B}\right)^\alpha = \mu_B{}^\alpha - \mu_A{}^\alpha \qquad (4.11)$$

Similarly we may show that

$$\left(\frac{dG}{dX_B}\right)^\beta = \mu_B{}^\beta - \mu_A{}^\beta \qquad (4.12)$$

Since at equilibrium, from (2.19), $\mu_B{}^\alpha = \mu_B{}^\beta$ and $\mu_A{}^\alpha = \mu_A{}^\beta$, then

$$\left(\frac{dG}{dX_B}\right)^\alpha = \left(\frac{dG}{dX_B}\right)^\beta \qquad (4.13)$$

Thus, for equilibrium, the slopes of the G^S-X_B curve at the compositions X^α and X^β must be equal. The only compositions for which this can be true for the curve in Fig. 4.3 is $X^\alpha = X^\beta = X^1$ and we are consequently led to the conclusion that any such mixture of two solutions in this system will at equilibrium become the single solution with composition equal to that of the alloy. The same argument can be extended to the possible splitting of, say, the solution X^β into still two other solutions, so that we conclude that the single solution is more stable than *any* other combination of solutions. This may also be seen by noting that the free energy of the single solution, G^1 in Fig. 4.3, is lower than that of any combination of solid solutions averaging to the composition X^1. For example, if the alloy consists of the phases α and β, as above, the free energy of the alloy per mole is, taking a weighted average of G^α and G^β,

$$G^{M\alpha\beta} = \frac{W^\alpha}{W^1} G^\alpha + \frac{W^\beta}{W^1} G^\beta \tag{4.14}$$

where W^α, W^β, and W^1 are the weights of the α and β phases and of the alloy as a whole, respectively. But

$$W^1 = W^\alpha + W^\beta \tag{4.15}$$

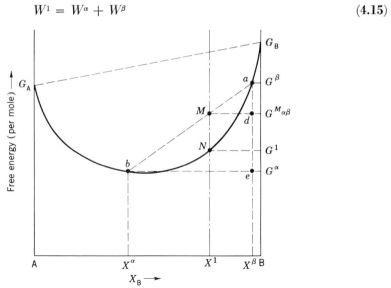

Fig. 4.3   *Free-energy curve for an ideal solution. The free energy of the single solution at N is lower than that for the mixture of solutions at M.*

and

$$X^1W^1 = X^\alpha W^\alpha + X^\beta W^\beta \tag{4.16}$$

so that combining (4.14) to (4.16),

$$G^{M\alpha\beta} - \frac{(X^\beta - X^1)G^\alpha + (X^1 - X^\alpha)G^\beta}{X^\beta - X^\alpha} \tag{4.17}$$

By adding and subtracting $X^\alpha G^\alpha$ in the numerator of (4.17) and rearranging, we obtain

$$G^{M\alpha\beta} = G^\alpha - (G^\alpha - G^\beta)\frac{X^1 - X^\alpha}{X^\beta - X^\alpha} \tag{4.18}$$

It may readily be seen that this value of $G^{M\alpha\beta}$ is given in Fig. 4.3 by the intersection, point M, of the composition vertical at X^1 and the chord ab, for by the principle of similar triangles

$$\frac{G^{M\alpha\beta} - G^\alpha}{G^\beta - G^\alpha} = \frac{X^1 - X^\alpha}{X^\beta - X^\alpha} \tag{4.19}$$

which is exactly equivalent to (4.18). Since M always lies above N in the figure, $G^{M\alpha\beta}$ is always greater than G^1; that is, the single solution is the stable phase!

The free-energy curve in Fig. 4.3 is given for some specific temperature. Since for any fixed composition, from Eq. (3.8),

$$\left(\frac{dG}{dT}\right)_P = -S \tag{3.8}$$

and since the entropy is positive, the free energy, and thus the curve in Fig. 4.3, is expected to decrease with increasing temperature. That is, curves similar to that in Fig. 4.3 but lying at different energy levels will represent the free-energy–versus–composition trend at other temperatures. This is illustrated in Fig. 4.4. It should be noted that, in accordance with Eq. (3.8), the free-energy curves in Fig. 4.4 are at lower levels with increasing temperature and are more sharply convex at high temperatures because of the increased importance of the $-T\Delta S_m$ term in the free-energy equation (4.8).

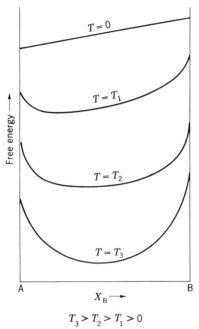

Free energy →

$T = 0$

$T = T_1$

$T = T_2$

$T = T_3$

A   B

$X_B \longrightarrow$

$T_3 > T_2 > T_1 > 0$

Fig. 4.4   *Free-energy curves for ideal solutions at several temperatures.*

4.5   EQUILIBRIUM PHASE DIAGRAM IN IDEAL SYSTEM

For a system in which ideal solutions are formed in both the liquid and solid states, the free-energy curves for both states will be of the form shown in Fig. 4.4. Consideration of the liquid and solid curves for several temperatures leads to the equilibrium phase diagram. A series of representative curves is depicted in Fig. 4.5. At a very high temperature, such as T_6 in Fig. 4.5, the liquid free-energy curve lies below that of the solid at all compositions; i.e., all alloys exist at equilibrium as a single liquid solution. With decreasing temperature the free energies of both liquid and solid increase in accordance with Eq. (3.8). Since, however, the entropy of a substance at a given temperature is generally higher in the liquid than in the solid state, the *rate of increase* of the liquid free energy is greater than that of the solid. Thus, the free-energy curve of the liquid in Fig. 4.5 moves up relative to that of the solid as the temperature decreases, and ultimately the two intersect. It is assumed in Fig. 4.5 that this happens first at pure A at the temperature T_5. Here the free energies of pure solid A and pure liquid A are equal; i.e., this is the melting temperature of pure A, the temperature at which pure liquid A and pure solid A may coexist in equilibrium. This is indicated by the point marked a at the temperature T_5 and the composition pure A in the temperature-com-

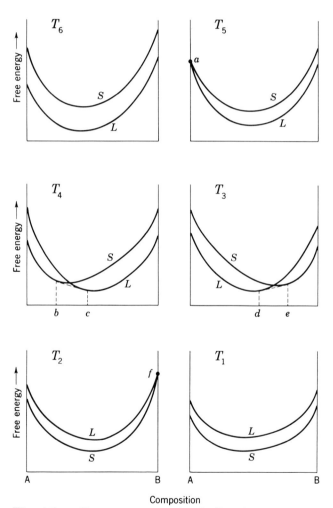

Fig. 4.5   *Free-energy curves for the liquid and the solid phases in an ideal system.*

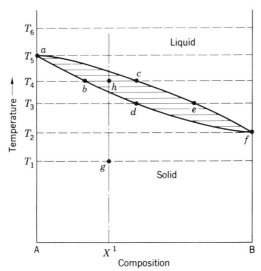

Fig. 4.6   *Ideal-system equilibrium phase diagram
resulting from the free-energy curves in Fig. 4.5.*

position phase diagram of Fig. 4.6. At all other compositions at T_5
the stable condition is a single liquid solution.

At T_4 the liquid and solid free-energy curves intersect at some
intermediate composition. Under these circumstances it may be
demonstrated that within a certain composition range a mixture
of two solutions, one liquid and one solid, is more stable than either
a liquid solution or a solid solution alone. Consider, for example,
the alloy of composition X^1 in the free-energy diagram of Fig. 4.7.
If this alloy exists as a single solid solution, it has the free energy
given by the point a in the diagram. It has the lower free energy
corresponding to point b if it exists as a single liquid solution, but
the lowest free energy it may have is given at point M, representing
the free energy of the alloy when it exists as a mixture of solid and
liquid solutions with compositions given at points d and e, the two
points of common tangency to the free-energy curves. That this
latter condition does in fact correspond to equilibrium may be seen
in the following way. Suppose the alloy X^1 exists as a mixture of
liquid and solid solutions of the general compositions X^L and X^S,
as shown in Fig. 4.7. Following Eq. (4.10), we may write for the
liquid phase

$$dG^L = \mu_A{}^L \, dX_A{}^L + \mu_B{}^L \, dX_B{}^L \qquad (4.20)$$

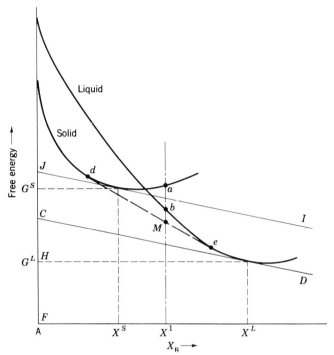

Fig. 4.7 *Intersecting free-energy curves; M is the lowest free energy in the alloy X^1.*

Since the partial molal free energy μ is by definition the free energy per mole of a component in solution, the free energy of the liquid solution G^L is

$$G^L = \mu_A{}^L X_A{}^L + \mu_B{}^L X_B{}^L \tag{4.21}$$

Upon substituting $X_A = 1 - X_B$ and $dX_A = -dX_B$, combining (4.20) and (4.21), and suitably rearranging, we obtain

$$\mu_B{}^L = G^L + (1 - X_B{}^L) \frac{dG}{dX_B{}^L}$$
$$\mu_A{}^L = G^L - X_B{}^L \frac{dG}{dX_B{}^L} \tag{4.22}$$

Now, comparing the second equation in (4.22) with the diagram in Fig. 4.7, it is seen that

$$G^L = \overline{HF}$$

and

$$\left(\frac{dG}{dX_{\mathrm{B}}}\right)^{L} = -\frac{\overline{CH}}{X_{\mathrm{B}}{}^{L}}$$

so that

$$\mu_{\mathrm{A}}{}^{L} = \overline{HF} + X_{\mathrm{B}}{}^{L}\frac{\overline{CH}}{X_{\mathrm{B}}{}^{L}} = \overline{HF} + \overline{CH} = \overline{CF}$$

In a similar way it may be shown that

$$\mu_{\mathrm{A}}{}^{S} = \overline{JF}$$

But at equilibrium the two chemical potentials are equal, and thus

$$\overline{CF} = \overline{JF} \tag{4.23}$$

which means that for equilibrium C and J represent the same point. This condition of equilibrium and that expressed in Eq. (4.13)

$$\left(\frac{dG}{dX_{\mathrm{B}}}\right)^{L} = \left(\frac{dG}{dX_{\mathrm{B}}}\right)^{S}$$

require, then, that the compositions X^{L} and X^{S} must at equilibrium be such that the tangents to the free-energy curves at these two compositions have the same slope [condition (4.13)] and a common point [condition (4.23)], in other words are the same straight line, the common tangent to the two curves.

It may be seen by inspection of Fig. 4.7 that this statement about the equilibrium condition of alloy X^{1} applies to all alloys the compositions of which lie between the points of common tangency d and e. (It is worth noting that this is also true regardless of the nature of the two phases involved, be they liquid-solid, liquid-liquid, or solid-solid.) Consequently, in all such alloys equilibrium at this temperature corresponds to a mixture of two phases, one liquid and one solid, and *regardless of the overall composition of these alloys the liquid phase always has the composition given by point e and the solid by point d* in Fig. 4.7. For alloys with overall composition to the left of point d, the lowest-free-energy condition is again to be found on the solid-solution curve, so that here the equilibrium condition is a

single solid solution. Similarly, to the right of point *e*, the stable
state is a single liquid solution.

Returning now to the free-energy diagram for the temperature
T_4 in Fig. 4.5, it is apparent that alloys of composition from pure A
to that corresponding to point *b* in the figure exist stably as a single
solid solution, those to the B-rich side of point *c* as a single liquid
solution, and those between *b* and *c* as a mixture of a solid solution
with composition *b* and a liquid solution with composition *c*. This
information is recorded in Fig. 4.6 by plotting the two compositions
b and *c* at the temperature T_4 on the temperature-composition phase
diagram. The horizontal isothermal line connecting the equilibrium
compositions *b* and *c* is called a *tie line*. Points *d* and *e* at the lower
temperature T_3 in Fig. 4.6 are obtained in a similar fashion from
Fig. 4.5.

As the temperature continues to decrease the liquid free-energy
curve ultimately will intersect that of the solid at pure B. The
temperature at which this happens, T_2 in Figs. 4.5 and 4.6, is the
melting point of pure B. Below this temperature the lowest free-
energy condition is a single solid solution at all compositions, as at
T_1. The equilibrium phase diagram is completed by drawing the
locus of all points representing compositions of liquid in equilibrium
with solid, the line *acef* in Fig. 4.6, and also the locus *abdf* of all
points representing the corresponding equilibrium solid compositions. Because the line *acef* gives the conditions above which all
alloys are completely liquid at equilibrium, it is called the *liquidus;*
similarly line *abdf*, giving the conditions below which all alloys are
solid, is called the *solidus*.

Although the phase diagram in Fig. 4.6 has been deduced for
essentially ideal systems, nonideal systems in which appreciable
but not excessive departures from ideality exist may also exhibit
this type of phase diagram if the departures from ideality are similar
in the liquid and solid states. Thus, it may be said that systems in
which virtual ideality exists will exhibit phase diagrams characterized by a lens-shaped liquid-solid region and complete solubility in
the liquid and solid states, but it is not necessarily true that the
existence of such a diagram guarantees ideality. Two examples of
systems with this simple diagram, the Cu–Ni and the Ge–Si systems,
are shown in Fig. 4.8.

The liquidus and solidus in such phase diagrams divide the
diagram into three regions, the one-phase liquid region, the one-
phase solid region, and the two-phase liquid-solid region in between.

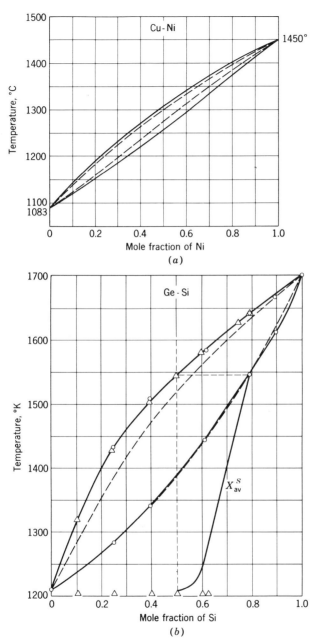

Fig. 4.8 (a) *The copper-nickel system. (From Seltz.[4.1]) (b) The germanium-silicon system. (From Thurmond.[4.2]) The full lines are experimental; the dashed lines were calculated from equations such as (4.35) and (4.36).*

It is important to realize that the one-phase regions have an entirely different significance and character than the two-phase regions. To illustrate, imagine that we make up an alloy of, say, composition X^1 in Fig. 4.6 by mixing together in a crucible the appropriate amounts of A and B. If we bring this alloy to equilibrium at the temperature T_1, it will be represented by the point g, which we will call its *alloy point*, in the one-phase solid region in the phase diagram. In the crucible we find a real physical entity, a phase, with the composition corresponding to g. On the other hand, if the alloy is equilibrated at T_4, it will be represented by the alloy point h, lying in the two-phase region, but there will now be two distinctly different physical entities in the crucible *neither* of *which has the composition h!* Thus, one-phase regions consist of an infinite number of alloy points each of which represents an attainable composition of a homogeneous, equilibrium phase; the alloy points in two-phase regions, however, do not correspond to any homogeneous entity actually attainable but merely indicate where in the diagram the compositions of the phases which actually do exist can be found. This distinction between one-phase regions and two-phase regions extends also to the multiphase regions which appear in more complex phase diagrams. In a sense, all multiphase regions in all phase diagrams can be looked upon as being simply the intersections between the various one-phase regions. In this view, only the one-phase regions have physical significance; the others are just constructions which give information about the various single phases coexisting under certain conditions.

4.6 EQUILIBRIUM SOLIDIFICATION IN SYSTEM WITH COMPLETE LIQUID AND SOLID SOLUBILITY

As an excercise in the interpretation of the diagram in Fig. 4.6, the equilibrium cooling of some typical alloy will be described in detail. Referring to Fig. 4.9, the alloy of composition X° exists as a single liquid solution of A and B at temperatures above T_1, the liquidus temperature for this composition. If heat is extracted at equilibrium rates, solidification begins at T_1. The solidus line gives at any temperature the composition of solid solution in equilibrium with liquid, and thus the composition of the first minute quantity of solid to form is $X_1{}^s$, given at the intersection of the T_1 temperature horizontal with the solidus line. Continued extraction

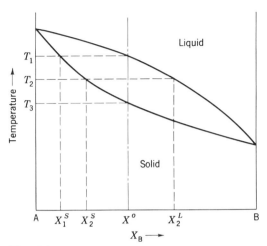

Fig. 4.9

of heat lowers the temperature; the liquid shifts its composition along the liquidus line, becoming more and more B-rich because of the rejection of A-rich solid. Each new increment of solid forms with the appropriate equilibrium composition given by the solidus at the temperature of formation; in addition, all the previously formed solid adjusts it composition to each new equilibrium value by exchange of A and B atoms across the solid-liquid interfaces. Thus, at any temperature such as T_2 in Fig. 4.9 all the liquid remaining has the composition X_2^L, and all the solid formed has the composition X_2^S. This process continues until, at T_3, where the composition of the solid is equal to that of the alloy, solidification is complete. Subsequent extraction of heat causes lowering of the temperature without further phase changes; the alloy remains a single solid solution. All other compositions in the system (except, of course, at $X = 0$ and $X = 1$) undergo exactly the same solidification process; upon heating, the sequence of events is reversed.

It should be noted that the solidification (and melting) in each two-component alloy takes place over a range of temperatures, whereas for the pure components A and B, freezing and melting are constant-temperature phenomena. These facts are directly predictable from the laws of thermodynamics as expressed in the phase rule and Le Chatelier's principle. The extraction of heat constitutes a constraint on the system in the form of a temperature gradient

at the boundary between the system and its surroundings.* The
system attempts to relieve this constraint; if, however, the system
consists of only one component, it cannot do so under equilibrium
conditions by lowering its temperature so long as both liquid and
solid are present, for according to the phase rule,

$$f = c - p + 1 = 1 - 2 + 1 = 0 \dagger$$

This means there is only one temperature at which the liquid and
solid can be in equilibrium. The system, therefore, relieves the constraint by transforming the high-heat-content phase, the liquid,
to the low-heat-content phase, the solid, at constant temperature.
On the other hand, when both components are present during
solidification,

$$f = c - p + 1 = 2 - 2 + 1 = 1$$

A degree of freedom is available; the system can, and does, therefore, lower its temperature to relieve the constraint represented by
the extraction of heat.

4.7   THE LEVER RULE

It is apparent from the preceding discussion that the equilibrium
phase diagram indicates the compositions of phases as a function
of temperature (and pressure) and the temperatures (and pressures)
of phase transitions as a function of composition. Another useful
type of information is also obtainable from the diagram, namely, the
amount of each phase present in fraction of the total alloy. In the
one-phase regions of, say, Fig. 4.9 the phase in question constitutes,
obviously, 100% of the alloy. In the two-phase region, the determination is also almost this simple. It is done by means of the
lever rule, which is simply a result of the simultaneous solution
of the two elementary weight balances in Eqs. (4.15) and (4.16).
Elimination of either W^L or W^S from these two equations as applied
to the temperature T_2 in Fig. 4.9 gives the fractional weights of

* The difference in temperature across the boundary is infinitesimal for equilibrium extraction of heat.
† One degree of freedom has been consumed in fixing the pressure at 1 atm;
thus the phase rule becomes $f = c - p + 1$ rather than $f = c - p + 2$.

the solid or the liquid phases, respectively, as

$$\frac{W^S}{W^\circ} = \frac{X_2{}^L - X^\circ}{X_2{}^L - X_2{}^S}$$

$$\frac{W^L}{W^\circ} = \frac{X^\circ - X_2{}^S}{X_2{}^L - X_2{}^S}$$

(4.24)

The latter equations are called the *lever rule* because they are the conditions of mechanical equilibrium which would apply to an imaginary lever along the tie line from $X_2{}^S$ to $X_2{}^L$ with its fulcrum at X°, the weight W^S at $X_2{}^S$, and the weight W^L at $X_2{}^L$.

4.8   NONEQUILIBRIUM SOLIDIFICATION IN SYSTEM WITH COMPLETE LIQUID AND SOLID SOLUBILITY. DIFFUSION

Strictly speaking, equilibrium cooling requires infinitesimally slow rates of heat extraction, i.e., rates not actually obtainable. Laboratory conditions, nevertheless, can frequently be arranged to give quite close approximations of equilibrium cooling rates. The major requirement is that sufficient time be allowed for the necessary continual adjustments in the phase compositions to occur. These adjustments take place by atom transfer across the interphase boundaries and by the movement of atoms from position to position within the phases, i.e., by *diffusion*. The ability of the atoms to diffuse is derived directly from the fact that at any finite temperature they have thermal kinetic energy, the average value of which is kT per atom, where k is the Boltzmann constant. In order to move from one position to another an atom must do so against an energy barrier Q, called the *activation energy*, which arises in large measure from the fact that the atom must push its way between surrounding atoms. In solids and, to a lesser extent in some liquids, Q is typically much larger than kT, and, hence, to make a successful movement an atom must acquire the energy Q by collisions with its neighbors. Statistical mechanics demonstrates that the probability of an atom's acquiring this energy is proportional to $\exp{(-Q/kT)}$ (the Boltzmann factor). Thus, it is clear that the rate of atom diffusion in any material should be strongly temperature-dependent; such rates are usually expressed in terms of a *diffusion coefficient D*, which experimentally is found to have the appropriate temperature dependence

$$D = D_0 \exp\left(-\frac{Q}{kT}\right)$$

where D_0 and Q are both constants specific to the particular substance concerned.

The values of D_0 and Q in most substances of interest to materials science (see, for example, Table 4.1) are such that in

Table 4.1   Some Values of Q and D_0 for Self-diffusion in Various Metals

Metal	D_0, cm²/sec	Q, kcal/mole	Ref.
SOLID:			
AG	0.40	44.1	4.4
AU	0.091	41.7	4.5
CU	0.20	47.1	4.6
γ-FE	0.58	67.9	4.7
LI	0.23	13.2	4.8
NA	0.24	10.5	4.9
NI	1.3	66.8	4.10
PB	0.28	24.2	4.11
LIQUID:			
HG	1.26×10^{-4}	1.16	4.12
NA	1.33×10^{-3}	2.58	4.13
PB	9.15×10^{-4}	4.45	4.14

practical situations cooling rates usually depart significantly from equilibrium. Under these conditions, the data plotted in an equilibrium diagram are no longer directly applicable. The diagram may, nevertheless, be used as a basis for understanding certain aspects of the departures from equilibrium to be expected. To illustrate this, let us consider the solidification of the alloy $X°$ in Fig. 4.9 under nonequilibrium conditions.* We shall assume for simplicity a cooling rate such that the time available at each temperature is

* For a more extensive discussion of nonequilibrium solidification see, for example, Ref. 4.3.

sufficient for virtually complete adjustment of the compositions within the liquid phase but for none within the solid phase. (This is essentially equivalent to the assumption that D in the solid is zero and D in the liquid is infinite.) Such a condition may be approached in practice, since, as shown in Table 4.1, values of Q in liquids are typically much smaller than those in solids; i.e., diffusion in the liquid form of a material may be expected to be orders of magnitude faster than that in the solid form at the same temperature (a factor of about 10^4 at the melting point).

Under these conditions the solidification of alloy $X°$ again starts at the temperature T_1 in Fig. 4.9, since the liquid is always in a condition of internal equilibrium. Also, the first minute quantity of solid to form may be supposed to have the composition $X_1{}^S$ since we may stipulate that at the liquid-solid interface equilibrium prevails in the solid as well as in the liquid. As cooling proceeds, the liquid composition shifts along the liquidus line and has reached $X_2{}^L$ at T_2. During this cooling new solid particles, or nuclei, will have formed, each with the composition appropriate to its formation temperature as given along the solidus line. Each of these nuclei, during the cooling subsequent to its formation, grows at the expense of the liquid. Both the newly formed nuclei and the growth increments on old nuclei will have compositions higher and higher in the component B as the temperature decreases. Since time is not available for the compositions within the solid to change once the solid has formed, at the temperature T_2 the solid will have compositions ranging all the way from $X_1{}^S$ to $X_2{}^S$, with an average somewhere between the two. This average will be represented by a point such as a in Fig. 4.10. Similar considerations reveal that at T_3 the solid composition will range from $X_1{}^S$ to $X_3{}^S$ with an average at a point such as b in Fig. 4.10. As the cooling continues, the average composition of the solid shifts toward higher B, until at T_4, the temperature at which the alloy $X°$ completes its solidification during *equilibrium* cooling, the average composition of the solid is at c in Fig. 4.10, where $c \neq X°$. It is apparent that in this case, the alloy cannot yet be completely solidified, for if an alloy consists of only one phase, the alloy composition and the phase composition must obviously be identical. Application of the lever rule also shows that there is still liquid remaining in the alloy at T_4, for the ratio of W^L to $W°$ here is

$$\frac{W_4{}^L}{W°} = \frac{c - X°}{c - X_4{}^L}$$

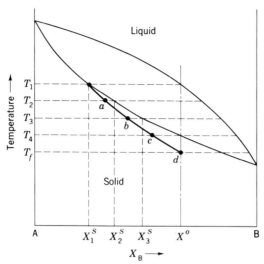

Fig. 4.10 *Composition (average) of the solid, line abcd, during nonequilibrium solidification of alloy $X°$.*

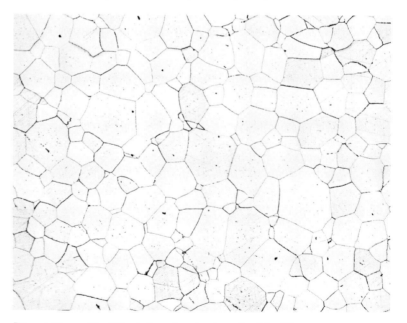

Fig. 4.11 *Equilibrated grain structure in Mo–W alloy containing 30% W. 75 ×. (From W. Rostoker and J. R. Dvorak, "Interpretation of Metallographic Structures," courtesy of Academic Press.)*

which is not zero. Solidification is not complete until the alloy reaches the temperature T_f, at which the line *abcd*, representing the average composition of the solid, crosses the alloy-composition line.

It is thus seen that nonequilibrium cooling of the type assumed leads to two major effects in the alloy. The first is an extension of the solidification temperature range with a resulting lowering of the final solidification temperature; if, as is usually the case, the liquid is also never quite at equilibrium, the temperature at which solidification begins will also be lowered somewhat. The second effect is the production of a solid with nonuniform composition. Because the solidification proceeds by the formation of numerous solid nuclei and the growth of these nuclei, each nucleus has a gradient of composition from its center, or core, to its periphery. As a result, this nonequilibrium effect is frequently referred to as *coring*. Illustrations of typical equilibrated and cored microstructures are shown in the photomicrographs of Figs. 4.11 and 4.12. Since each of the solid nuclei formed during cooling grows until it

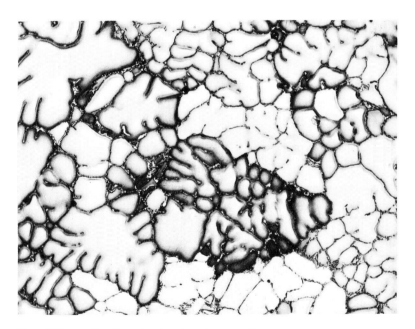

Fig. 4.12   *Cored grain structure in as-cast copper-in-aluminum alloy. Keller's etch. 100×. (Courtesy of T. R. Pritchett and R. A. Ridout, Kaiser Aluminum and Chemical Sales Corp.)*

impinges on similarly growing neighbors, there results in the completely solidified material an array of *grains*, or *crystallites*, separated by *grain boundaries*. The boundaries are simply regions of misfit, where the more or less randomly oriented crystal lattices of the various grains meet. In an alloy solidified under virtually equilibrium conditions, or in one subsequently equilibrated by reheating, the grains are uniform in composition and thus in appearance, as shown in Fig. 4.11. In a rapidly cooled alloy (Fig. 4.12), the coring is evidenced by the gradations in shading from center to periphery of the grains.

4.9   ZONE REFINING

A further instance of nonequilibrium cooling is of particular interest because of its relationship to the very useful process of purification called *zone refining*. To study this kind of nonequilibrium cooling let us examine solidification in a binary alloy under the following conditions:

1 The cooling rate is such that negligible diffusion takes place in the solid; in the liquid, diffusive mixing does take place but at a finite rate.

2 There is no convective mixing in the liquid.

3 The solute concentrations in the liquid X^L and in the solid X^S at any given temperature are maintained at equilibrium at the liquid-solid interface.   The equilibrium ratio of X^S/X^L is designated k, the *distribution coefficient*.

Since zone refining is concerned primarily with very low concentrations of solute, let us further limit our attention to dilute solutions and for simplicity make the approximation that the liquidus and solidus in this range are straight lines. As will be seen in Chaps. 6 and 7, k can be either less than or greater than unity. We shall discuss the case for $k < 1$ (the principles are not altered for $k > 1$). When a small solid nucleus forms at T_1 (Fig. 4.13) in a liquid of initial composition $X°$, the solid will have the composition $kX°$, and solute will be rejected by the newly formed solid into the surrounding liquid. This excess solute will be carried away by diffusion in the liquid, but since both D in the liquid and the rate of advance of the solid-liquid interface as the nucleus grows are finite, the concentration X^L in the liquid near the interface will remain higher than $X°$. As the temperature drops to the liquidus temperature corresponding to the composition X^L, the interface advances

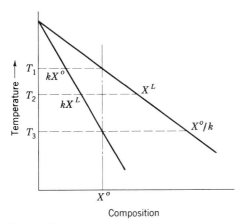

Fig. 4.13

into the surrounding liquid, the newly forming solid now having the composition $kX^L > kX°$. This process continues, and at some temperature, such as T_2 in Fig. 4.13, the composition profile from the center of the nucleus across the interface into the liquid can be represented as in Fig. 4.14a. Both X^S and X^L in Fig. 4.14a will continue to increase until $X^S = X°$ and $X^L = X°/k$ at the temperature where the alloy-composition vertical and the solidus line cross (T_3 in Fig. 4.13). Here the conditions for steady-state solidification are met, and no further changes in X^S and X^L take place (until the growing nucleus approaches another nucleus). This can be demonstrated with the help of Fig. 4.14b. To simplify this argument, the diffusion profile in the liquid from X^L to $X°$ is approximated as a straight line. When the interface of the solid nucleus advances the small distance dy, the liquid-concentration profile moves from position ABC to position DEF. If steady state prevails, this means that the profile does not change; i.e., the total amount of solute, the area under the profile, remains constant. This in turn means that the amount of solute rejected in forming the solid (area $ABED$) is equal to the amount (area $BEFC$) which diffuses down the concentration gradient in raising each point on the gradient profile to its new value. But

$$ABED = (X^L - X^S)\ dy$$

and

$$BEFC = \overline{EF}\ dz$$

Also by similar triangles

$$\overline{EF}\ dz = \overline{BH}\ dy = (X^L - X^\circ)\ dy$$

and therefore

$$(X^L - X^S)\ dy = (X^L - X^\circ)\ dy$$

or, at steady state

$$X^S = X^\circ$$

Steady-state solidification prevails until the interface approaches another growing nucleus or the end of the sample. At this point the general composition of the liquid beyond point F in Fig. 4.14b rises because there is no longer a constant-composition "sink"

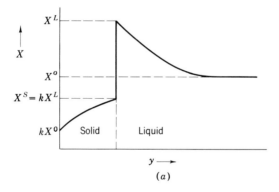

(a)

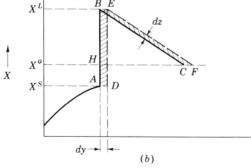

(b)

Fig. 4.14 *Composition profiles in the solid and liquid during unidirectional, nonequilibrium freezing.*

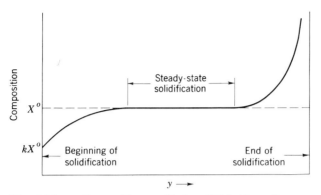

Fig. 4.15   *Composition profile in solidified bar after unidirectional, nonequilibrium solidification.*

into which to drain the solute rejected from the solid. When this happens, X^L, and therefore X^S, both increase again, leading to a profile in the solid as indicated in Figure 4.15.

It may be seen that at the start of solidification the solid has a much lower solute content than did the liquid from which it formed. This fact is taken advantage of in zone refining, first developed by Pfann,[4.15] in the following way. A sample of the impure material is prepared in the form of a long bar. A short molten zone is then passed from one end of this bar to the other (Fig. 4.16). This is

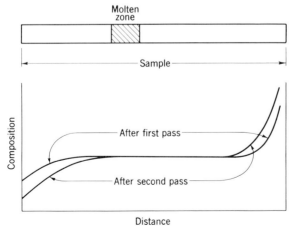

Fig. 4.16   *Composition profiles after first and second zone-refining passes.*

done, for example, by moving a small annular heater along the bar. After one pass the melting at the leading end of the zone and solidification at the trailing end produce a concentration profile for the impurities having $k < 1$ similar to that in Fig. 4.15. Purification is accomplished for a short distance at the start, the impurities having been passed to the finishing end of the bar. A second pass

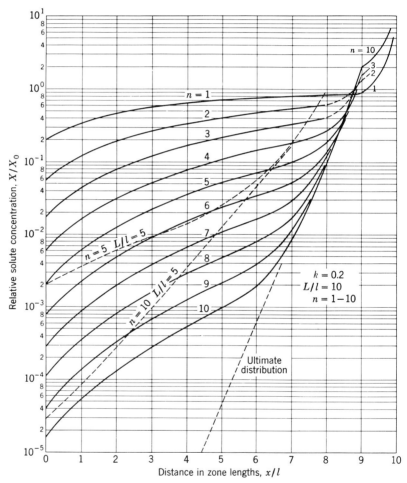

Fig. 4.17   *Logarithm of relative solute concentrations $X/X°$ versus distance in zone lengths x/l from beginning of bar in zone refining, for various numbers of passes n; L denotes bar length. (From Pfann.[4.15])*

superimposes a similar redistribution of solute on the profile resulting from the first pass (see Fig. 4.16), and so on. A short molten
zone is used to avoid the general mixing which would result if
after each pass the whole bar were remelted as a unit. After several
passes a very high degree of purification can be obtained in part of
the sample. The concentration profile in the solid after one pass is
relatively easy to calculate and is given for all but the final zone
length by the equation

$$\frac{X}{X^\circ} = 1 - \left[(1 - k) \exp\left(-\frac{ky}{l} \right) \right]$$

where X° is the initial concentration, l is the zone length, and X is
the concentration in the solid at the distance y from the starting
end. Multiple-pass distributions are not readily represented by
equations; they have, however, been presented graphically for
various conditions, the curves being based on approximate equations.[4.15] A typical example is given in Fig. 4.17. Of course, it should
be emphasized that during each pass, impurities for which $k > 1$
move countercurrent to the molten zone; i.e., they are removed
from the finishing end of the bar and piled up at the starting end.
Thus in practice, the middle portion of the bar will have the highest
purity if both types of impurities are present.

The advent of zone refining has increased by several orders of
magnitude the purities with which materials can be prepared. In
some materials, impurity contents as low as 10^{-10} atomic fraction
have been produced. This has had important repercussions in all
areas of materials technology; perhaps the most significant of these
have been in electronics, where a whole new industry based on semi-
conductors was developed only when zone refining made possible
the production of semiconductor materials of sufficient purity.

4.10   CALCULATION OF LIQUIDUS AND SOLIDUS IN
IDEAL SYSTEM

Real systems which behave almost ideally in both the solid and
liquid states present the possibility of the relatively simple quantitative calculation of phase diagrams. As an example of the ultimate
potential of the thermodynamic method in this respect the equa-

tions for the liquidus and solidus curves in such systems will be developed from the relations previously derived in this chapter.

From Eq. (4.8) we may write

$$G^L = G^{M,L} + RT(X_A{}^L \ln X_A{}^L + X_B{}^L \ln X_B{}^L) \tag{4.25}$$

where G^L is the free energy per mole of the liquid solution having the mole fractions $X_A{}^L$ and $X_B{}^L$ at the temperature T and $G^{M,L}$ is the free energy per mole which a mixture of undissolved pure liquid A and pure liquid B in the same proportions would have at T. $G^{M,L}$ in turn may be written

$$G^{M,L} = X_A{}^L G_A{}^L + X_B{}^L G_B{}^L \tag{4.26}$$

where $G_A{}^L$ is the molal free energy of pure liquid A and $G_B{}^L$ that of pure liquid B at T. The chemical potential of B in the solution at equilibrium is, from Eq. (4.22),

$$\mu_B{}^L = G^L + X_A{}^L \frac{dG^L}{dX_B} \tag{4.27}$$

Differentiating Eqs. (4.25) and (4.26) gives

$$\frac{dG^L}{dX_B} = G_B{}^L - G_A{}^L + RT \ln \frac{X_B{}^L}{X_A{}^L} \tag{4.28}$$

and substitution of Eqs. (4.25), (4.26), and (4.28) into (4.27) yields, with appropriate rearrangement,

$$\mu_B{}^L = G_B{}^L + RT \ln X_B{}^L \tag{4.29}$$

In an analogous manner, it may be seen that

$$\mu_B{}^S = G_B{}^S + RT \ln X_B{}^S \tag{4.30}$$

where $G_B{}^S$ is the free energy per mole of pure solid B at the temperature T and $X_B{}^S$ is the mole fraction of B in the solid solution at T. If the liquid and solid solutions are to be in equilibrium, the chemical potentials of B in each must be identical; equating (4.29) and (4.30),

$$RT \ln \frac{X_B{}^S}{X_B{}^L} = G_B{}^L - G_B{}^S = \Delta G_B \tag{4.31}$$

ΔG_B is the free energy of fusion for pure B at the temperature T, which may be expressed in terms of the more commonly available quantity $\Delta H_B{}^m$, the enthalpy of fusion of B at its melting temperature $T_B{}^m$, in the following way. At a given temperature

$$\Delta G_B = \Delta H_B - T \Delta S_B \tag{4.32}$$

When B melts under equilibrium conditions at constant temperature and pressure, i.e., at $T_B{}^m$, the change in free energy is zero, according to Eq. (2.12), so that

$$\Delta G_B{}^m = 0 = \Delta H_B{}^m - T_B{}^m \Delta S_B{}^m$$

or

$$\Delta S_B{}^m = \frac{\Delta H_B{}^m}{T_B{}^m} \tag{4.33}$$

Assuming that neither ΔS_B nor ΔH_B varies with temperature,[*] i.e., that $\Delta S_B = \Delta S_B{}^m$ and $\Delta H_B = \Delta H_B{}^m$, then substitution of (4.33) into (4.32) gives

$$\Delta G_B = \Delta H_B{}^m - T \frac{\Delta H_B{}^m}{T_B{}^m} = \Delta H_B{}^m \left(1 - \frac{T}{T_B{}^m} \right) \tag{4.34}$$

Finally, by combining (4.31) and (4.34) we find

$$\ln \frac{X_B{}^S}{X_B{}^L} = \frac{\Delta H_B{}^m}{R} \left(\frac{1}{T} - \frac{1}{T_B{}^m} \right) \tag{4.35}$$

The same analysis obviously holds for the component A in the two solutions, so that

$$\ln \frac{1 - X_B{}^S}{1 - X_B{}^L} = \frac{\Delta H_A{}^m}{R} \left(\frac{1}{T} - \frac{1}{T_A{}^m} \right) \tag{4.36}$$

Thus, a knowledge of the equilibrium melting temperatures and the enthalpies of fusion (equal to the heats of fusion at constant pressure) of the pure components at their melting points allows simultaneous solution of the two Eqs. (4.35) and (4.36) for the two unknowns $X_B{}^S$ and $X_B{}^L$ at each temperature, yielding the equilib-

[*] This is equivalent to assuming that the difference between the heat capacity of the liquid and that of the solid, ΔC_P, does not change with T, a good approximation for the relatively narrow range of T of interest here.

rium phase diagram. Examples of such calculated curves are given
in Fig. 4.8 for the systems Cu–Ni and Ge–Si.

4.11   THE FREE ENERGY OF NONIDEAL SOLUTIONS

Real solutions are, of course, not ideal. The number of different
ways the free energy of real solutions may depart from ideality is
large, but fortunately the number of distinctly different types of
two-component phase equilibria resulting is relatively small. Thus,
we shall cover here only some of the simpler types of departures
from ideality and the corresponding equilibrium diagrams.

According to Eqs. (4.1) to (4.3), the free energy of a solution,
if the atomic mixing is random, is

$$G^S = G^M + X_A(\bar{H}_A - H_A) + X_B(\bar{H}_B - H_B)$$
$$- T[X_A(\bar{S}_A - S_A) + X_B(\bar{S}_B - S_B)]$$
$$+ RT(X_A \ln X_A + X_B \ln X_B) \quad (4.37)$$

or, grouping the various quantities,

$$G^S = G^M + \Delta H^{xs} - T \Delta S^{xs} - T \Delta S_m$$
$$= G^M + \Delta G^{xs} - T \Delta S_m \quad (4.38)$$

Comparing (4.38) with (4.7), it may be noted that for an ideal
solution $\Delta G^{xs} = \Delta G^{xs,id} = 0$, and thus ΔG^{xs} for any real solution
is the difference between the actual free energy of the solution and
the value the free energy would have if the solution were ideal.
ΔG^{xs} is consequently designated the excess free energy of solution;
similarly ΔH^{xs} is the excess enthalpy, and ΔS^{xs} the excess entropy,
of solution. When ΔG^{xs} is positive, a solution is said to have a
positive deviation from ideality. Conversely a negative ΔG^{xs} corresponds to a negative deviation from ideality.

It is clear from Eqs. (4.37) and (4.38) that the nature of the
constant-pressure–constant-temperature free-energy curve of a nonideal solution as compared with that of an ideal solution (see Fig.
4.2) is complicated by the presence of a finite ΔG^{xs}. Since in general
the magnitude of ΔG^{xs} and its variation with composition, temperature, and pressure is as yet not calculable from theory but must
usually be obtained by experiment, it follows that the shape of the
free-energy–composition curve is also not generally predictable, as

was the case for the relatively simple ideal-solution curve of Fig. 4.2. By making certain simplifying assumptions, however, useful information can be obtained on how ΔG^{xs} is expected to affect equilibrium diagrams.*

Let us assume first that ΔS^{xs} is negligibly small, which is equivalent to limiting the discussion to approximately *regular* solutions, a regular solution being defined as one for which $\Delta S^{xs} = 0$. Equation (4.38) then becomes

$$G^S = G^M + \Delta H^{xs} - T\,\Delta S_m \qquad (4.39)$$

This condition is approached in many liquids and in some close-packed solids such as metals. The difference between the free-energy curve of such a solution and that of an ideal solution lies entirely in the term ΔH^{xs}. To get some idea of how this term may be expected to affect the free-energy curve, it is instructive to invoke an atomistic point of view which has come to be called the *quasi-chemical* approach, invented by Guggenheim[4.16] in 1935 for the calculation of the enthalpies of regular solutions.

4.12   QUASI-CHEMICAL CALCULATION OF THE EXCESS ENTHALPY OF SOLUTION

From Eqs. (4.37) and (4.38)

$$\Delta H^{xs} = H^S - H^M$$

Since $H = E + PV$, and since the internal energy is made up of the sum of the potential energies between the atoms E^{PE} and their thermal kinetic energies E^{KE}, we may write

$$\Delta H^{xs} = (E^{PE} + E^{KE} + PV)^S - (E^{PE} + E^{KE} + PV)^M \quad (4.40)$$

For condensed phases PV is small at ordinary pressures, and also generally $V^S - V^M$ is small, so that $PV^S - PV^M \cong 0$. The kinetic energies are also virtually unaffected by composition and so $E^{KE,S} \cong E^{KE,M}$. Equation (4.40) thus becomes

$$\Delta H^{xs} = E^{PE,S} - E^{PE,S} \qquad (4.41)$$

* See Chap. 8 for an example of the direct calculation of a phase diagram of intermediate complexity where ΔG^{xs} is calculated on the basis of certain physically measurable parameters and simple assumptions about ΔG^{xs} versus X.

In the quasi-chemical approach it is now assumed that an expression for E^{PE} can be obtained by assigning an interaction energy $\mathcal{V}_i$ to each type of atom pair in the system and summing over all atom pairs; that is,

$$E^{PE} = \Sigma n_i \mathcal{V}_i$$

where n_i is the number of pairs having interaction energy $\mathcal{V}_i$. The nearest-neighbor approximation is generally applied; i.e., it is usually assumed that only nearest-neighbor interactions need be counted. (This actually accounts for about 80 to 90% of the total interaction energy; the other 10 to 20% is often not important qualitatively but can be very important if the attempt is made to use the theory quantitatively.)

Let us now apply this concept to the calculation of $E^{PE,S}$ in a two-component solid solution of atoms A and B having a crystal structure with the coordination number Z. In this solution there are three types of nearest-neighbor pairs, AA, BB, and AB pairs. To find the number of AB pairs, n_{AB}, consider the surroundings of, say, any A atom. It has Z nearest neighbors, and if the atoms are assumed to be intermixed at random, the fraction X_B of these will be B atoms. Thus, each A atom has ZX_B B neighbors. In a mole of solution there are $X_A N$ (N = Avogadro's number) A atoms, so that the total number of AB pairs is

$$n_{AB} = ZNX_A X_B \tag{4.42}$$

Each A atom also has ZX_A A neighbors, thus

$$n_{AA} = \frac{ZN}{2} X_A{}^2 \tag{4.43}$$

the factor $\frac{1}{2}$ being introduced since in this case each AA pair was counted twice, once for each A atom in it. Similarly

$$n_{BB} = \frac{ZN}{2} X_B{}^2 \tag{4.44}$$

The interaction energy $E^{PE,S}$ is, then,

$$E^{PE,S} = \frac{ZN}{2} X_A{}^2 \mathcal{V}_{AA} + \frac{ZN}{2} X_B{}^2 \mathcal{V}_{BB} + ZNX_A X_B \mathcal{V}_{AB} \tag{4.45}$$

The energy $E^{\mathrm{PE},M}$ of a mechanical mixture of pure A and pure B is found in like manner to be

$$E^{\mathrm{PE},M} = \frac{ZN}{2} X_A \upsilon_{AA} + \frac{ZN}{2} X_B \upsilon_{BB} \qquad (4.46)$$

Combining (4.41), (4.45), and (4.46), we see that

$$\Delta H^{\mathrm{xs}} = ZNX_A X_B \left(\upsilon_{AB} - \frac{\upsilon_{AA} + \upsilon_{BB}}{2} \right)$$

or

$$\Delta H^{\mathrm{xs}} = ZNX_A X_B \upsilon \qquad (4.47)$$

where

$$\upsilon = \upsilon_{AB} - \frac{\upsilon_{AA} + \upsilon_{BB}}{2} \qquad (4.48)$$

An expression similar in form can in principle be developed for liquids; the expression would differ chiefly in that neither Z nor the interatomic distances between nearest neighbors would be exactly identifiable—average values would have to be used instead.

In regular solutions, ΔH^{xs} constitutes the deviation of the solution from ideality; it is now seen that in the quasi-chemical view the direction of this deviation depends on the sign of υ, which in turn depends on the relative magnitudes of υ_{AB}, υ_{AA}, and υ_{BB}. If υ_{AB} is equal to the average of υ_{AA} and υ_{BB}, that is, if the attraction between unlike atoms is the same as that between like atoms, then $\Delta H^{\mathrm{xs}} = 0$, and the solution is ideal. If like atoms attract more strongly than unlike atoms (υ_{AB} greater than the average of υ_{AA} and υ_{BB}*), ΔH^{xs} is positive; i.e., there is a positive deviation from ideality. This corresponds to a situation in which there is a tendency for the separation of A-rich and B-rich phases; the solution is only maintained when the opposing tendency toward complete randomness—the thermal agitation as expressed in the entropy of mixing—is stronger. Since the latter effect decreases in intensity as the temperature decreases, ΔH^{xs} positive generally indicates a tendency toward limited solubility at low temperatures. Conversely, negative ΔH^{xs} (υ_{AB} less than the average of υ_{AA} and

---

* Attractive forces are negative, so that stronger attraction between like atoms means υ_{AB} is a *smaller negative* number than the average of υ_{AA} and υ_{BB}; that is, it is *greater* than $(\upsilon_{AA} + \upsilon_{BB})/2$.

$\mathcal{V}_{BB}$) means unlike atoms have the stronger attractions in solution; in this case there is no tendency for the separation of A-rich and B-rich phases even at low temperatures. On the contrary, decreasing the thermal energy of the atoms in the solution tends to stabilize an *ordered* condition in which the probabilities of like and unlike atom pairs in the solution are nonrandom, nearest-neighbor pairs tending to be made up of unlike atoms. In liquids, and in solids at high temperatures, the prevailing high thermal energies limit the non-random correlations between atom pairs to pair distances of only a few atomic diameters. This is called *short-range order*. In solids at low temperatures, however, the ordering tendency frequently leads to the extension of these correlations over regions many thousands of atom diameters in breadth, i.e., to *long-range order*. The ordering phenomenon is discussed in detail in Chap. 5.

4.13   FREE-ENERGY CURVES OF REGULAR SOLUTIONS BASED ON QUASI-CHEMICAL CALCULATION OF ΔH^{xs}

From Eq. (4.47) it is evident that at constant temperature ΔH^{xs} is a symmetrical parabolic function of composition having the value zero at $X_A = 0$ or $X_B = 0$ and a maximum or minimum (depending on the sign of $\mathcal{V}$) at $X_A = X_B = 0.5$. (This assumes that $\mathcal{V}$ is not a function of composition, an assumption which is justified only as a first approximation.) When ΔH^{xs} is negative, the constant-temperature free-energy curves have the same simple convex-down-

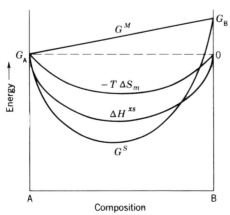

Fig. 4.18   *Energy-composition curves for* ΔH^{xs} *negative.*

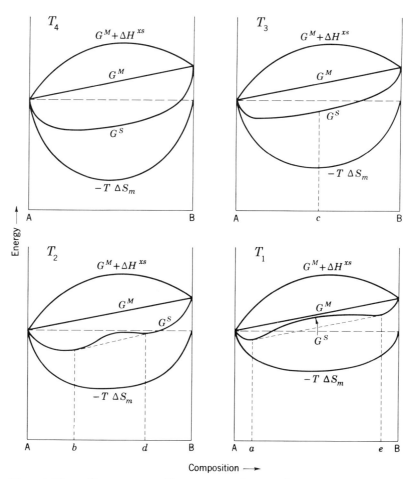

Fig. 4.19   *Energy-composition curves at several temperatures for ΔH^{xs} positive. $T_4 > T_3 > T_2 > T_1$.*

ward shape as those for ideal solutions. This is illustrated in Fig. 4.18. The lowest-free-energy condition is the single solution at all compositions, just as in the case of ideal solutions. Here, however, the tendency for ordering at low temperatures exists.

When ΔH^{xs} is positive, the situation is somewhat more complicated (Fig. 4.19). As shown in Sec. 4.3, the slope of the G^s-X curve must always be negative at and near the composition extremities. At low temperatures, however, the maximum negative magnitude of the entropy term relative to the maximum positive magnitude of the enthalpy term is such that the G^s-X curve inflects

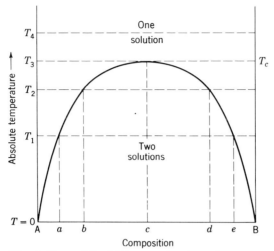

Fig. 4.20   *Miscibility gap in phase diagram corresponding to free-energy curves in Fig. 4.19.*

in the central composition range (T_1 and T_2, Fig. 4.19). As the temperature is raised and the entropy term becomes more important, the inflection in the free-energy curve disappears; the temperature at which this just takes place (T_3 in Fig. 4.19) is frequently called the *critical temperature* T_c. At higher temperatures (T_4 in Fig. 4.19) the free-energy curve of the solution again has the same simple shape as for ΔH^{xs} zero or negative. When an inflection is present in the free-energy curve, the lowest-free-energy conditions at various alloy compositions are determined with the help of the common tangent to the two convex-downward portions of the curve, just as in the case of the intersecting free-energy curves discussed in connection with Fig. 4.7. This means that for alloy compositions between the two points of common tangency a mixture of two solutions with compositions corresponding to the points of common tangency is the stable (lowest-free-energy) condition; for other alloy compositions a single solution is stable. In the solid portion of the equilibrium phase diagram this leads to a two-solutions region, or *miscibility gap*, as shown schematically in Fig. 4.20.

Liquid miscibility gaps are not to be expected in systems of moderate ΔG^{xs}, for the entropy-of-mixing terms are normally high enough to eliminate the inflection in the G^S-X curves at the lowest temperatures of liquid stability. Liquid miscibility gaps do occur quite frequently in other systems (see Chaps. 6 and 7).

4.14   EQUILIBRIUM DIAGRAMS FOR NONIDEAL ISOMORPHOUS SYSTEMS OF MODERATE ΔG^{xs}; ΔG^{xs} POSITIVE

It is clear from the discussion in the preceding section that the nature of the phase diagram characterizing a given system depends on the sign of ΔG^{xs} (ΔH^{xs} in regular solutions). The nature of the diagram also depends on the magnitude of ΔG^{xs}. Very high positive values of

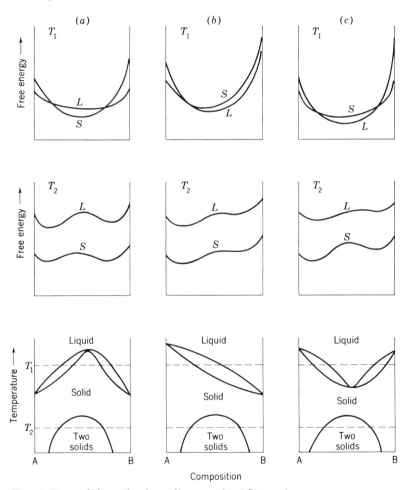

Fig. 4.21   *Schematic phase diagrams for ΔG^{xs} positive. (a) $\Delta G^{xs,L}$ more positive than $\Delta G^{xs,S}$; (b) $\Delta G^{xs,L} \cong \Delta G^{xs,S}$; (c) $\Delta G^{xs,L}$ less positive than $\Delta G^{xs,S}$.*

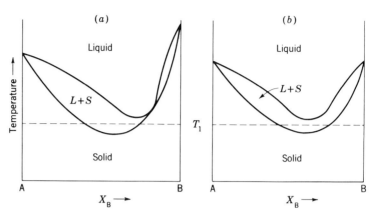

Fig. 4.22   *Two incorrect constructions for diagrams having minima in solidus and liquidus curves. (a) Compositions of solid and liquid not the same at minima; (b) solidus and liquidus not touching at minima.*

ΔG^{xs} tend to be associated with highly immiscible components, i.e., with relatively pure *terminal** phases in a system, whereas high negative values of ΔG^{xs} tend to be associated with the existence of *intermediate** (compoundlike) phases.

The relative magnitudes of $\Delta G^{xs,L}$ and $\Delta G^{xs,S}$, the excess free energy of solution in the liquid as compared with that in the solid state, also affect the nature of a phase diagram. For isomorphous systems with moderate values of ΔG^{xs}, Fig. 4.21 depicts the three types of diagrams to be expected when ΔG^{xs} is positive. In each of these a miscibility gap is shown at low temperatures, although in practice such gaps are only infrequently found because of the slow diffusion rates and the consequent difficulty in attaining equilibrium at low temperatures. Figure 4.21a shows the development of a maximum in the liquid-solid region corresponding to the condition $\Delta G^{xs,L}$ more positive than $\Delta G^{xs,S}$. When $\Delta G^{xs,L} \cong \Delta G^{xs,S}$ the diagram is as shown in Figure 4.21b with the lens-shaped liquid-solid region similar to that discussed in connection with ideal systems. Finally, when $\Delta G^{xs,L}$ is less positive than $\Delta G^{xs,S}$ the diagram exhibits a liquid-solid region with a minimum, as in Fig. 4.21c. It is worth emphasizing that whenever the liquidus and solidus

---

* A *terminal* phase is defined as one which includes in its compositional range of stability the mole fraction 1 or 0; an *intermediate* phase includes neither of these compositions.

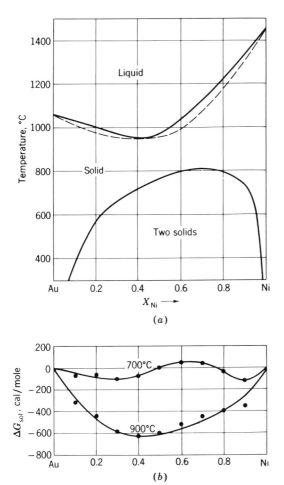

Fig. 4.23   (a) *The gold-nickel equilibrium phase
diagram. (From Hansen.*[4.17]) (b) *The free energy of
solution for solid gold-nickel solutions at 700 and
900°C. (Calculated from data from Seigle, Cohen, and
Averbach.*[4.19])

curves in a diagram exhibit minima (or maxima), the minima (or
maxima) must occur at the same composition, and the two curves
must touch at this composition. That this is so may be deduced
from a series of free-energy curves such as those in Fig. 4.21; it may
also be seen by reference to the two incorrect constructions in Fig.
4.22. In Fig. 4.22*a* the solidus and liquidus have different composi-

tions at the minima, and in Fig. 4.22*b* they do not touch at the minima; in both instances it is apparent that there is a range of temperatures, including that marked T_1, where isothermal lines in the liquid-plus-solid region do not intersect the liquidus. That this situation is thermodynamically impossible is demonstrated by the nature of the free-energy curves which give rise to the liquidus and solidus, curves such as those in the upper diagram of Fig. 4.21*c*. At any temperature the region where two phases may coexist is delineated by the two points of common tangency on the free-energy curves, one on the liquid curve and one on the solid. Since these two points must also delineate the two-phase region in the phase diagram (see Figs. 4.5 and 4.6), it is clear that each horizontal line, or tie line, in such a region must have one end on the liquidus and the other on the solidus.

Examples of real systems having the forms shown in Fig. 4.21 are given in Figs. 4.23 to 4.26. The Au–Ni and the LiCl–NaCl systems (Figs. 4.23 and 4.24) illustrate the liquidus-solidus minimum type in Fig. 4.21*c*; the Au–Pt and the ThO_2–ZrO_2 systems (Figs. 4.25 and 4.26) illustrate the lens-shaped liquid-solid type of

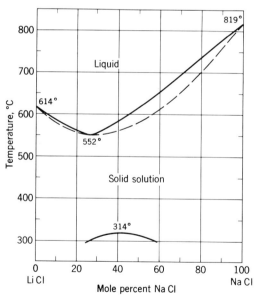

Fig. 4.24   *The lithium chloride–sodium chloride equilibrium phase diagram. (From Levin, McMurdie, and Hall.*[4.18]*)*

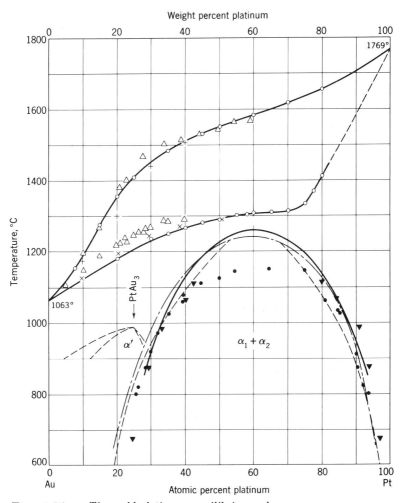

Weight percent platinum

Fig. 4.25   *The gold-platinum equilibrium phase
diagram. (From Hansen.[4.17])*

Fig. 4.21*b*. No examples of the configuration in Fig. 4.21*a* have been
found. Included in Fig. 4.23 are curves for the free energy of solid
solution ΔG_{sol} as a function of composition in the Au–Ni system at
700 and 900°C. In terms of the quantities used in Eq. (4.38)

$$\Delta G_{\text{sol}} = G^S - G^M = \Delta G^{\text{xs}} - T \, \Delta S_m$$

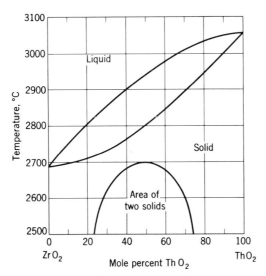

Fig. 4.26   *The thorium oxide–zirconium oxide (ThO_2–ZrO_2) equilibrium phase diagram; probable diagram. (From Levin, McMurdie, and Hall.[4.18])*

where all quantities here refer to solids. It may be seen that the measured thermodynamic data predict the type of phase diagram actually found; in fact, the equilibrium solubility values which may be deduced by drawing the common tangent to the 700°C free-energy curve are in reasonably good quantitative agreement with the solubility limits shown in the phase diagram.

4.15   EQUILIBRIUM AND NONEQUILIBRIUM COOLING THROUGH THE MISCIBILITY GAP. THE SPINODAL

From an equilibrium point of view, the two-phase region within the solid-solution miscibility gap in Fig. 4.21a, b, or c has exactly the same characteristics as that between the liquidus and solidus. When a liquid solution is cooled to the liquidus line, the liquid phase becomes saturated with respect to a solid of composition given at the intersection of the temperature tie line with the solidus. In like manner, when a solid solution is cooled to the *solvus* line, or boundary of the two-solids region, the solid solution becomes saturated with respect to a second solid solution. For example, the alloy of composition $X°$ in Fig. 4.27 exists as a single solid solution during cooling through the temperature range from the solidus to the temperature T_1. At the latter temperature, however, it becomes saturated with respect to a second solid solution having a composition given on the solvus at b, the intersection of the T_1-temperature tie line with the solvus. For identification purposes this second solid solution is now called β and the original solution α. Thus, upon extraction of an infinitesimal quantity of heat at T_1, the alloy consists of the parent solid solution of composition $X°$ and a minute quantity of the new solution of composition $X_1{}^\beta$. Continued extraction of heat produces a lowering of the temperature, just as it does during cooling through the liquid-plus-solid region, and further formation of the solid solution β from the parent solution α. If the cooling is carried out at equilibrium (or near-equilibrium) rates, the composition of the parent solution shifts along the solvus toward higher A content, and the composition of the newly forming β and all the previously formed β shifts along the solvus toward higher B content. At T_2 the α solid solution will have attained the composition $X_2{}^\alpha$ and the β the composition $X_2{}^\beta$. The average composition of the alloy is, of course, still $X°$, and application of the lever rule reveals that at T_2 the fraction $f_2{}^\beta$ of β solid solution in the alloy is

$$f_2{}^\beta = \frac{X° - X_2{}^\alpha}{X_2{}^\beta - X_2{}^\alpha}$$

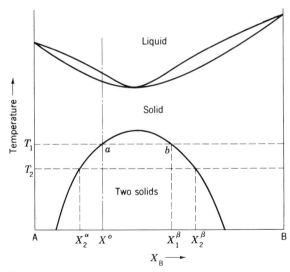

Fig. 4.27

Nonequilibrium cooling through the miscibility gap gives rise to the interesting concept of the spinodal curve. This curve is obtained from free-energy–composition curves such as those in the lower diagrams of Fig. 4.19. It will be seen that there are two inflection points in each such curve; the spinodal is the locus of all the inflection points plotted in the temperature-composition phase diagram, as shown schematically in Fig. 4.28. The significance of the spinodal may be outlined as follows. Although in equilibrium cooling of an alloy through the miscibility gap the separation of the second solution starts at the solvus line, in nonequilibrium cooling a finite amount of undercooling is necessary to obtain appreciable rates of separation. To understand why this is so one must visualize the details of the process by which a nucleus of the new phase forms. Consider an alloy of composition X° cooled to some temperature such as T_1 (Fig. 4.28). In order for a nucleus of ultimate composition X^β to form, a small group of atoms, an *embryo*, must acquire a higher B content than the surroundings, and an interface must appear. It can be shown that when such a group of atoms is both very small and has not yet raised its B content very much above the average, the presence of the interface and the composition change are accompanied by an *increase* in overall free energy rather than by the *decrease* associated with the final formation of a large particle of composition X^β. Only when this embryo has grown beyond a certain *critical size* and *critical composition* is its further growth associated with a decrease in free energy. (In most of the

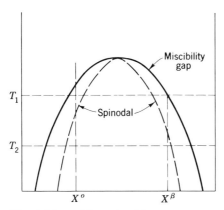

Fig. 4.28   *The spinodal curve.*

considerations in this book it is implicitly assumed that the phases dealt with have their equilibrium concentration when formed and are of large size, i.e., large compared to atomic dimensions.) We may thus say the embryonic nucleus must surmount both a *size barrier* and a *compositional barrier* before it can grow stably.

The size barrier is a consequence of the fact that energy must always be *supplied* to form the interface. This energy is provided by the decrease in bulk free energy accompanying the formation of the new phase, and since the surface-to-volume ratio of an embryo decreases as the embryo grows in size, the negative change in bulk free energy on forming the embryo outstrips the positive change due to the interface formation only above some critical size. This size barrier is present at all temperatures; it becomes smaller as the temperature of transformation decreases, however, because the specific surface energy of the interface does not vary much with temperature, but the bulk-free-energy decrease becomes greater at lower transformation temperatures. As the barrier to the transformation becomes smaller, the rate of the transformation increases, and thus the new phase forms at observable rates only after some finite degree of undercooling.

The compositional barrier also contributes to the need for a finite degree of undercooling to obtain observable rates of transformation. This barrier is present, however, only at temperatures above the spinodal, as may be demonstrated by reference to the free-energy curves in Fig. 4.29. When the B-rich embryo forms, it leaves a B-poor "halo" in the immediately adjacent parent solution. At a temperature such as T_1 in Fig. 4.28, where the undercooled solution is *above* the spinodal, the free-energy relationships are as shown in Fig. 4.29a, b, and c. Here it is seen that the average free energy of the embryo-plus-halo region must first rise (point M in Fig. 4.29a) as the embryo composition $X^{\beta'}$ moves toward the equilibrium value X^β. Only when, by chance fluctuations, the embryo composition reaches that shown in Fig. 4.29b does the average free energy drop on further embryo enrichment, as indicated by point M in Fig. 4.29c. The increase in free energy accompanying the early compositional growth of the embryo constitutes a compositional barrier which lowers the rate of phase separation. However, at temperatures *below* the spinodal, such as T_2 in Fig. 4.28, the bulk-free-energy change on forming an embryo-plus-halo region is negative for even the smallest possible difference in composition between embryo and parent solution. This is indicated by the point M in

Fig. 4.29d. Thus, below the spinodal the compositional barrier disappears, and only the size barrier restrains phase separation. As a result, transformation in solutions quenched to temperatures somewhat below the spinodal is considerably less sluggish than in those quenched to temperatures somewhat above the spinodal.

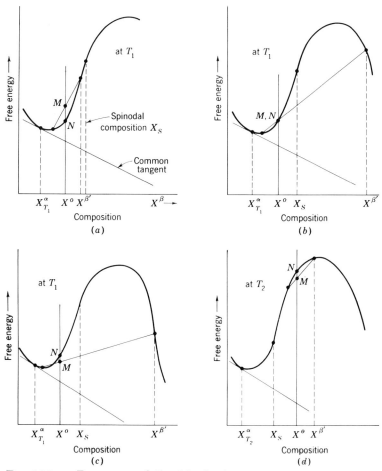

Fig. 4.29 *Free-energy relationships in phase separation above the spinodal (a to c), and below the spinodal (d). The free-energy change MN on forming an embryo of composition $X^{\beta'}$ is positive, then negative at T above spinodal (T_1); MN is always negative at T below spinodal (T_2).*

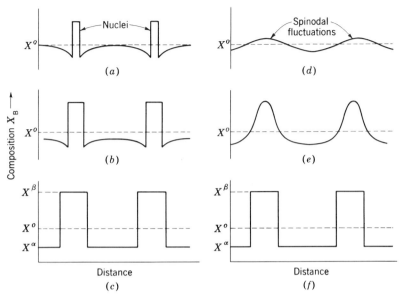

Fig. 4.30   *Schematic composition profiles during successive stages of phase separation by (a to c) nucleation and growth and by (d to f) spinodal decomposition.*

The morphology of the transformation product changes also in the spinodal region. Above the spinodal, the large compositional difference necessary for stable growth of the new phase ensures that there will be a discrete boundary between the phases, i.e., discrete nuclei will form, even in the early stages of decomposition. This is illustrated schematically in the composition plot of Fig. 4.30a. Below the spinodal, on the other hand, even regions of very small compositional difference from the matrix can grow stably provided they are large enough in extent; thus, in the early decompositional stages, the boundary regions between the phases will no longer be sharp: there will simply be a gradual and periodic composition variation throughout the parent solution with quite diffuse boundaries between the high and low composition regions. This is depicted schematically in Fig. 4.30d. In the later stages of the transformation, the appearance of the spinodal decomposition product will approach that characteristic of discrete particles and the two modes of transformation will then not be readily distinguishable (Fig. 4.30b, c, e, and f).

4.16   QUASI-CHEMICAL CALCULATION OF THE
MISCIBILITY GAP AND THE SPINODAL

Using the quasi-chemical treatment, an approximate calculation of the position of the two-solids miscibility gap for very simple, regular systems can be made. From Eqs. (4.39) and (4.47)

$$G^S = G^M + ZNX_A X_B \upsilon + RT(X_A \ln X_A + X_B \ln X_B)$$

Minimizing G^S with respect to, say, X_B (and noting that

$$X_A = 1 - X_B)$$

we obtain

$$\frac{dG^S}{dX_B} = 0 = \frac{dG^M}{dX_B} + ZN\upsilon(1 - 2X_B) + RT \ln \frac{X_B}{1 - X_B} \quad (4.49)$$

If it is now assumed that the two pure components are sufficiently alike so that $G_A \cong G_B$ *, then

$$\frac{dG^M}{dX_B} \cong 0$$

and the minima in the free-energy curve will coincide with the points of common tangency. Solution of (4.49) then gives for the equilibrium compositions

$$\frac{RT}{NZ\upsilon} = \frac{1 - 2X_B}{\ln\left[(1 - X_B)/X_B\right]} \quad (4.50)$$

For the condition $G_A \cong G_B$, this is a symmetrical function in X having a maximum at $X = 0.5$ and $T = T_c$, the critical temperature. It contains only one unknown constant, υ. Solving for υ in terms of T_c, we find

$$\upsilon = \frac{2RT_c}{NZ} \quad (4.51)$$

---

* This is equivalent to assuming that $\upsilon_{AA} \cong \upsilon_{BB}$, and that the specific heats of the pure substances are also approximately equal.

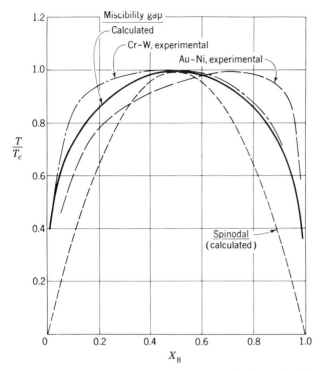

Fig. 4.31   *The miscibility gap and the spinodal according to Eqs. (4.52) and (4.55), respectively, and the experimental curves for the Cr–W and the Au–Ni systems. For the experimental curves, W and Ni are taken as component B. (Data from Hansen[4.17])*

and inserting this into (4.50) gives

$$\frac{2(1 - 2X_B)}{\ln\left[(1 - X_B)/X_B\right]} = \frac{T}{T_c} \tag{4.52}$$

A plot of Eq. (4.52) is presented in Fig. 4.31 along with two typical curves determined experimentally for the systems Cr–W and Au–Ni. The curve for Cr–W fits the calculated curve quite well (perhaps fortuitously so), that for the Au–Ni system not so well. The discrepancies are primarily a result of the many simplifying assumptions made in deriving Eq. (4.52), in particular the assumptions that υ is not a function of composition and that $G_A \cong G_B$. In addition, the especially large discrepancies between the calculated curve and that for the Au–Ni system may be partly a result of a basic inadequacy in the quasi-chemical concept. Thermodynamic measurements on the Au–Ni system[4.19] have indicated that ΔH^{xs} in this system is indeed positive in the temperature range of the solubility gap but that this is the result of strain energy arising from the size difference between the Au and Ni atoms rather than a result of the favoring of like interaction energies between atom pairs. In fact, some slight tendency toward short-range ordering, indicating the favoring of unlike nearest-neighbor bonds, was found at temperatures just above the solubility gap! Thus, strain energies, which are always positive, can play an important part in determining phase diagrams. The quasi-chemical concept, however, does not take this into account, since these strain energies are not readily represented as pair interactions but result rather from a general distortion of the crystal structure. The existence of such strain energies undoubtedly helps account for the fact that in the approximately 800 metal-system phase-diagrams shown in Hansen's 1958 edition of the "Constitution of Binary Alloys" not one example of the type of diagram shown in Fig. 4.21a is to be found. This type of diagram, with a liquidus-solidus maximum and a solid-solution miscibility gap, requires ΔG^{xs} positive and $\Delta G^{xs,L} > \Delta G^{xs,S}$. Since, however, strain energies are positive and can only appear in solids, their existence lowers considerably the probability of finding $\Delta G^{xs,L} > \Delta G^{xs,S}$.

An approximate calculation of the spinodal curve can be made with the same assumptions as those which led to Eq. (4.52) for the miscibility gap. Thus, differentiating Eq. (4.49),

$$\frac{d^2G^S}{dX_B{}^2} = -2N\upsilon Z + RT\,\frac{1}{X_B(1 - X_B)} \tag{4.53}$$

Setting this equal to zero gives the inflection point in Eq. (4.49),

$$\frac{RT}{2N\upsilon Z} = X_B(1 - X_B) \tag{4.54}$$

and solving for υ in terms of T_c ($T = T_c$ where $X_B = 0.5$) and substituting into Eq. (4.54) gives

$$4X_B(1 - X_B) = \frac{T}{T_c} \tag{4.55}$$

This is the quasi-chemical equation of the spinodal; it is plotted a s
a dashed line in Fig. 4.31.

4.17   EQUILIBRIUM DIAGRAMS FOR NONIDEAL
ISOMORPHOUS SYSTEMS OF MODERATE ΔG^{xs}; ΔG^{xs}
NEGATIVE

Based on the energy-composition curves in Fig. 4.18, the three
diagram types shown in Fig. 4.32 may generally be expected in
isomorphous systems of moderate, negative ΔG^{xs}. The free-energy
curves for both liquid and solid are convex downward at all temperatures, and the particular liquidus-solidus shape exhibited again
depends on whether $\Delta G^{xs,L}$ is more negative than $\Delta G^{xs,S}$ (Fig. 4.32a),
approximately equal to $\Delta G^{xs,S}$ (Fig. 4.32b), or less negative than
$\Delta G^{xs,S}$ (Fig. 4.32c). It may be seen by comparison with Fig. 4.21
that all three types of liquid-solid regions may result from either
ΔG^{xs} positive or ΔG^{xs} negative. These liquidus-solidus shapes cannot,

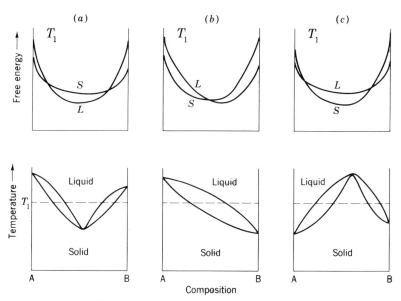

Fig. 4.32   *Schematic phase diagrams for* ΔG^{xs}
negative. (a) $\Delta G^{xs,L}$ *more negative than* $\Delta G^{xs,S}$; *(b)*
$\Delta G^{xs,L} \cong \Delta G^{xs,S}$; *(c)* $\Delta G^{xs,L}$ *less negative than* $\Delta G^{xs,S}$.

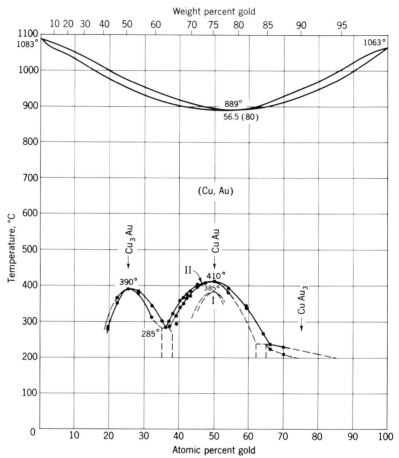

Fig. 4.33 *The gold-copper equilibrium phase diagram. (From Hansen.[4.17])*

therefore, be used to predict the sign of ΔG^{xs} (or ΔH^{xs}). In the case of ΔG^{xs} negative, however, the diagrams do not display the miscibility gap but rather tend to exhibit ordering or intermediate-phase formation at low temperatures. Examples of actual systems are given in Figs. 4.33 to 4.35, representing the Au–Cu, Cu–Pd, and Cu–Pt systems. For comparison with Figs. 4.18 and 4.32, measured values of ΔH^{xs}, $-T\,\Delta S = -T(\Delta S_m + \Delta S^{xs})$, and $\Delta G = \Delta G^{xs} - T\,\Delta S_m$ for the solution of solid copper and gold at 750°C are shown in Fig. 4.36.

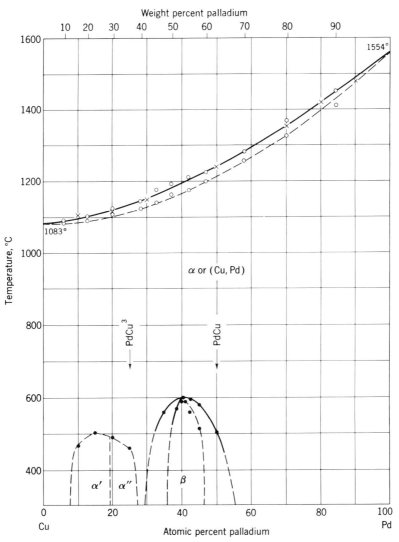

Fig. 4.34   *The copper-palladium equilibrium phase diagram. (From Hansen.[4.17])*

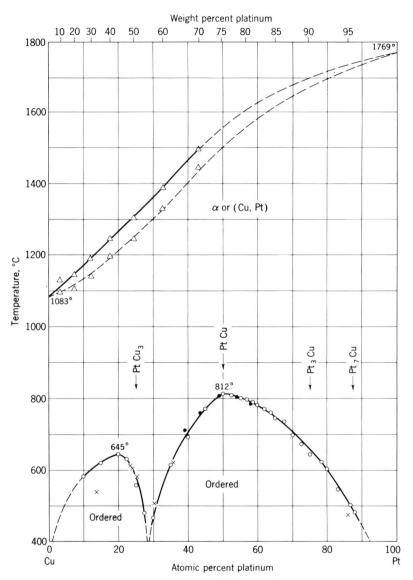

Fig. 4.35 *The copper-platinum equilibrium phase diagram. (From Hansen.[4.17])*

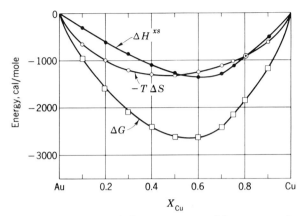

Fig. 4.36   *The enthalpy, entropy, and free energy of solution for solid copper and gold at 750°C. (Data from Kubaschewski and Catterall.[4.20])*

The ordering transformation is discussed in the next chapter, the formation of intermediate phases in Chap. 6.

Problems

4.1   From the ideal-gas law, $PV = nRT$, derive the expression for the ideal entropy of mixing, $\Delta S_m = -R(X_A \ln X_A + X_B \ln X_B)$.

4.2   Assume that you wish to calculate the internal energy of a phase in a phase mixture to an accuracy of $\pm 10\%$. If the phase is present as small particles, if the surface layer of atoms on each particle is assumed to be two atomic diameters thick, and if the internal energy per unit volume of the surface material is twice that of the bulk material, calculate approximately the smallest particle size for which the internal-energy calculation may be made neglecting the difference in energy of the surface and bulk material.

4.3   Equation (4.5) demonstrates thermodynamically that absolutely pure substances are not stable. This means, for example, that a simple binary eutectic diagram (see Fig. 6.4 for eutectic diagram) showing no terminal solid solubilities is impossible. Demonstrate this latter statement phenomenologically (rather than thermodynamically) by consideration of the melting of the pure components and of the eutectic mixture in such a system.

4.4   Plot $(\Delta G^{id,S} - \Delta G^M)/RT$ from Eq. (4.8) with special attention to small values of X_B and X_A.

4.5   (a) Making the assumption that Eqs. (4.35) and (4.36) apply to the Au–Pt system, calculate the liquidus and solidus in this system using the following data:

	T_m, °K	ΔH_m, cal/mole
GOLD	1336	2955
PLATINUM	2042	4700

(b) Compare with the Au–Pt phase diagram in Fig. 4.25. From the agreement, or lack thereof, between the calculated and experimental lines, what can you say as to the ideality of Au–Pt solutions? (c) What does the nature of the remaining portion of the experimentally determined diagram indicate about the ideality or departure therefrom of the system?

4.6   For equilibrium cooling of a 50 atomic percent Ge–Si alloy, what fraction of the alloy would be solid at 1500°C? (See Fig. 4.8.)

4.7   In the discussion of nonequilibrium solidification on page 67, it is assumed that the liquid adjusts its composition completely to the equilibrium values whereas the solid undergoes no adjustment. To check the reasonableness of this assumption calculate the relative penetration distances of a solute in liquid lead as compared to that in solid lead, assuming the pertinent diffusion rates have the same ratio as that given for liquid and solid lead in Table 4.1. The penetration distance may be taken as $d = 2\sqrt{Dt}$, where t is the time for penetration to the distance d.

4.8   Two of the major impurities in commercial "high-purity" aluminum are copper and silicon. If a bar of such aluminum, initially containing 0.01 weight percent of each, is zone-refined, calculate the weight percent of each impurity at a position five zone lengths from the starting end of the bar after 1 pass and after 10 passes. Use the tables on pp. 211–230 of Ref. 4.15, and any necessary phase-diagram data from published diagrams. Take L/l (Ref. 4.15) to be 10.

4.9   Assuming the Au–Pt system forms regular solid solutions, and using the quasi-chemical approximation, calculate and compare the magnitudes of the two quantities ΔH^{xs} and $T\,\Delta S_m$ for the 50 atomic percent alloy in this system (a) at 1200°C, and (b) at 700°C.

4.10   Assuming that the Au–Ni solid solutions are regular solutions, (a) calculate ΔH^{xs} for the 70% Ni alloy at 900°C from the free-energy data in Fig. 4.23; (b) calculate the same quantity from the phase diagram in Fig. 4.23 and Eqs. (4.47) and (4.51); (c) from a comparison of these two values for ΔH^{xs} for the 70% Ni alloy at 900°C, discuss the probability that these Au–Ni solutions are in fact regular solutions.

4.11   Cook and Hilliard [*Trans. AIME*, **233**:142 (1965)] have shown that in systems where the miscibility gap meets the composition of the pure components at 0°K, the spinodal may be calculated from the phase diagram by the equation

$$X_s - X_c \cong (X_e - X_c)\left(1 - 0.422\,\frac{T}{T_c}\right)$$

where X_c is the critical composition and X_s and X_e are the spinodal and equilibrium compositions, respectively, at the temperature T. Calculate the spinodal for the Cr–W system using this equation. Plot this curve, the regular solution spinodal (Fig. 4.31), and the experimental miscibility gap.

References

4.1   H. Seltz, *J. Am. Chem. Soc.*, **56**:307 (1934).

4.2   C. D. Thurmond, *J. Phys. Chem.*, **57**:827 (1953).

4.3   B. Chalmers, "Physical Metallurgy," chap. 6, John Wiley & Sons, Inc., New York, 1959.

4.4   C. Tomizuka and E. Sonder, *Phys. Rev.*, **103**:1182 (1956).

4.5   S. Makin, A. Rowe, and A. LeClaire, *Proc. Phys. Soc. (London)*, **70B**:595 (1957).

4.6   A. Kuper, H. Letaw, L. Slifkin, and C. Tomizuka, *Phys. Rev.*, **98**:1870 (1955).

4.7   F. S. Buffington, I. D. Bakalor, and M. Cohen, Self-diffusion in Iron, in "Physics of Powder Metallurgy," McGraw-Hill Book Company, New York, 1951.

4.8   C. P. Slichter, Diffusion Effects in Magnetic Resonance of the Alkali Metals, in "Defects in Crystalline Solids," Physical Society, London, 1955.

4.9   N. Nachtrieb, E. Calalano, and J. A. Weil, *J. Chem. Phys.*, **20**:1185 (1952).

4.10   R. Hoffman, F. Pickur, and R. Ward, *Trans. AIME*, **206**:483 (1956).

4.11   N. Nachtrieb and G. S. Handler, *J. Chem. Phys.*, **23**:1569 (1955).

4.12   R. Hoffman, *J. Chem. Phys.*, **20**:1567 (1952).

4.13   N. Nachtrieb and G. S. Handler, *Acta Met.*, **2**:797 (1954).

4.14   S. J. Rothman and L. D. Hall, *Trans. AIME*, **206**:199 (1956).

4.15   W. G. Pfann, "Zone Melting," John Wiley & Sons, Inc., New York, 1958.

4.16   E. A. Guggenheim, *Proc. Roy. Soc. (London)*, **A148**:304 (1935).

4.17 M. Hansen, "Constitution of Binary Alloys," 2d ed., McGraw-Hill Book Company, New York, 1958.

4.18 E. M. Levin, H. F. McMurdie, and F. P. Hall, "Phase Diagrams for Ceramists," The American Ceramic Society, Columbus, Ohio, 1956.

4.19 L. L. Seigle, M. Cohen, and B. L. Averbach, *Trans. AIME*, **194**:1320 (1952).

4.20 O. Kubaschewski and J. A. Catterall, "Thermochemical Data of Alloys." Pergamon Press, New York, 1956.

FIVE

THE ORDER-DISORDER TRANSFORMATION*

5.1   THE NATURE OF ORDERING

There are many types of order in solid materials, e.g., the spatial order of the atoms in a pure crystal; the term ordering in substitutional solid solutions, however, is usually reserved for the special atomic arrangements characterized by higher than random probabilities of finding unlike atoms as nearest neighbors. We are primarily concerned with the relationship of the ordering phenomenon to phase diagrams; since, however, certain general aspects of ordering seem to be difficult for students to grasp, we shall first consider in some detail the concept of ordering and some physical models proposed to allow its mathematical treatment. The quasi-chemical mathematical treatment which predicts the phase-diagram relationships will then be described.

Experimentally, two kinds of order have been detected and identified in solid solutions, *short-range order* and *long-range order*. Both are the result of the tendency of each atom in certain solutions to surround itself with unlike atoms. When this tendency is small,

* The author is indebted to L. Guttman, Argonne National Laboratories, for considerable help in clarifying the ideas expressed here and for the development of much of the mathematics associated with the quasi-chemical model outlined in this chapter.

or when at high temperatures the disordering tendency of thermal agitation is high, any specific atom succeeds in surrounding itself with only a slightly greater than random number of unlike atoms and for only very short times. Thermal agitation breaks up these local groups, and new groups form, which are in turn "dissolved," etc. Nevertheless, if we were to "freeze in" an atomic distribution at any moment and count the number of like and unlike nearest-neighbor pairs we should find the fractional number (the probability) of unlike pairs to be higher than that for a random distribution. The probabilities of unlike next-nearest pairs, third-nearest pairs, etc., would also be found to deviate from randomness, but the magnitude of the deviation would drop off rapidly with distance, most often falling virtually to zero at a few atomic diameters. Ordering of this type is called short-range ordering.

If the ordering tendency is high, or if the temperature is low, the nonrandom correlations between the positions occupied by the two atomic species are higher in magnitude and are effective over greater distances. Below some *critical temperature* the distance becomes large enough so that time-independent, long-distance correlations, long-range order, appear. In this state it is possible to classify all the lattice sites in the crystal into *sublattices*, each of which tends to be occupied predominantly by one kind of atom (see Fig. 5.1). If, when long-range order exists, we again count pairs of atoms, we shall now find that not only do nearest-neighbor and next-nearest-neighbor pairs exhibit nonrandom probabilities of like and unlike pairs but that there are equally prominent and nonvanishing correlations between the states of occupancy of sites that are separated by distances approaching macroscopic dimensions.

There is actually a sharper distinction between short-range and long-range order than may at first glance seem to be the case. The presence of long-range order and the magnitude of the critical temperature are clearly defined both theoretically and experimentally. Theoretically, as will be seen later, long-range order is either definitely present or definitely not present, depending on whether the temperature is above or below the critical temperature; the theoretically predicted position of the latter is, however, a function of the type of assumptions made in deriving the relationship between temperature and the ordering tendency. Experimentally, the presence of long-range order is delineated primarily by certain characteristic x-ray diffraction effects. This may be understood with reference to Fig. 5.1, which shows the completely ordered arrangement in a 50:50 A–B alloy having a bcc space lattice in the disordered

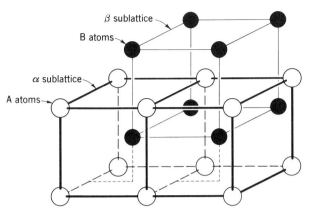

Fig. 5.1   *The ordered arrangement in a 50:50 alloy which has a bcc space lattice in the disordered condition. If the atomic diameters are equal, the ordered space lattice is simple cubic with an AB pair associated with each lattice point. The α and β sublattices are indicated.*

state. This ordered structure may be viewed as being made up of two interpenetrating sublattices, on one of which, called the α sublattice, the A atoms reside and on the other, the β sublattice, the B atoms reside. Each α site is surrounded by eight equidistant β sites, and similarly each β site by eight equidistant α sites. If this ordered arrangement of the atoms extends over thousands of atom diameters, it represents perfect long-range order. It should be noted that in the perfectly ordered condition the space lattice is no longer bcc, since the positions occupied by the A atoms no longer have surroundings equivalent to those occupied by the B atoms. In fact, the structure shown in Fig. 5.1 is characterized by a simple cubic space lattice with an AB atom pair associated with each lattice point.

As may be seen, long-range order lowers the symmetry of the crystal structure; as a result, it is detectable by the appearance in x-ray diffraction patterns of new reflections, called *superlattice* reflections. The intensities of these reflections are directly related to the *degree of long-range order*, i.e., to the extent of separation of the atomic species onto the appropriate sublattices. An example of the appearance of superlattice lines in x-ray powder patterns is shown in Fig. 5.2. Short-range order alone does not produce superlattices or superlattice reflections but rather affects the intensity of the diffuse (background) scattering in x-ray diffraction patterns, and is measurable by this effect.

The characteristics of short- and long-range order may be further illustrated by noting the following points:

1 Perfect short-range order implies perfect long-range order, and vice versa.

2 Completely random arrangements of the two kinds of atoms are characterized by no short-range and no long-range order.

3 Intermediate arrangements, however, can have quite high degrees of short-range order and little or no long-range order.

An extreme case of this is illustrated in Fig. 5.3 for a hypothetical 50:50 A–B alloy made up of out-of-step blocks, or *antiphase domains.* It is seen that in this two-dimensional alloy the α and β sublattices over long distances are each occupied 50% by A and 50% by B atoms, and therefore there is no long-range order in the alloy. On the other hand, it is also apparent that the arrangement is quite highly ordered on a local scale, the number of unlike nearest-neighbor atom pairs being far greater than for a random arrangement. The number for a random arrangement is given by Eq. (4.42) as

$$n_{AB}{}^r = ZNX_AX_B \tag{4.42}$$

In Fig. 5.3, $Z = 4$ for the 48 interior atoms, 2 for the 4 corner

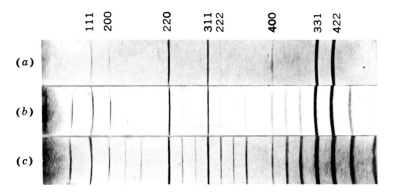

Fig. 5.2   *X-ray powder-diffraction patterns of the superlattice* $AuCu_3$. *(a) Disordered; (b) partially ordered; (c) highly ordered.* [*From Barrett and Massalski,*[5.1] *originally taken from C. Sykes and H. Evans, J. Inst. Met.,* **58**: 255 *(1936).*]

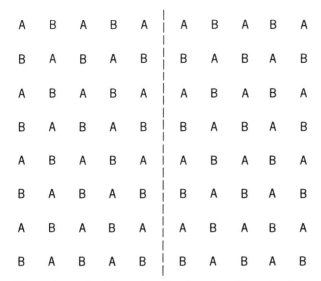

Fig. 5.3   *Antiphase domains in a two-dimensional crystal. The broken line, an antiphase domain boundary, separates two perfectly ordered domains which are out of step with each other.*

atoms, and 3 for the 28 other exterior atoms, and $X_A = X_B = \frac{1}{2}$, giving $n_{AB}{}^r = \frac{284}{4} = 71$. The actual number of AB pairs in the figure is 134, and the total number of pairs is 142. Thus, it is seen that the fractional number of AB pairs in the alloy would be 0.5 for a random array whereas in the actual array it is 0.95.

5.2   PHYSICAL MODELS OF ORDERING

Because the ordering tendency in alloys is opposed by thermal agitation, there is for each temperature a characteristic equilibrium degree of order at which the two opposing tendencies are just balanced. It is the derivation of the relation between this equilibrium degree of order and temperature which leads to theoretical order-disorder lines in phase diagrams. Two major steps are involved in this derivation: (1) visualizing an appropriate physical model of ordering and (2) devising a mathematical treatment of the model amenable to solution. This has turned out to be a formidable task, so much so that after 40 years of attack neither a completely satisfactory model nor a more than approximately solvable mathematical treatment has been found.

The most frequently and successfully used physical model of ordering was first proposed by Bethe[5.2] in 1935. This model is the exact analogy of the quasi-chemical model introduced for regular solutions by Guggenheim[4.16] at about the same time (see page 79). Since the quasi-chemical model seems more natural and is of more general use, we shall from now on refer to this concept as the quasi-chemical (Q-C) model. According to the model, the ordering is the result of nearest-neighbor atom-pair interaction energies for which the attractive interaction between unlike atoms is greater than the average of that between like atoms. This view has led to important qualitative, and some semiquantitative, agreement between theory and experiment. It should be noted, nevertheless, that it has a number of serious shortcomings. Among these are the neglect of such factors as the interactions between non-nearest-neighbor atoms, the distortion of lattice symmetry and size during the order-disorder transitions, and the changes in vibrational entropy accompanying the transition. In addition it cannot explain the existence, in a few instances, of ordered structures with very large numbers of atoms in their unit cells; models[5.3,5.4] based on the electronic structure of the crystal as a whole have been proposed to account for these cases. There is little doubt, also, that ordering may be induced in solutions of atoms with appropriately different diameters by the tendency to minimize the strain energy attendant on the misfit distortion of the crystal structure; such a tendency cannot readily be accounted for in terms of nearest-neighbor pair-interaction energies. In spite of these shortcomings, the Q-C model has been the starting point for more successful treatments of ordering than any other model and thus will be used here as an instructional base.

The earliest model leading to a moderately successful mathematical treatment for the stability of superlattices as a function of temperature was devised by Bragg and Williams in 1934.[5.5] Their model proposed that ordering is the result of the long-range influence of all the atoms in a crystal on each individual atom. They derived their order-temperature relation without reference to pair interactions. The Bragg-Williams (B-W) result, however, can be obtained on the basis of the Q-C model, as well, simply by applying to the pair-interaction concept assumptions equivalent to those made in the B-W treatment.

5.3   ORDER PARAMETERS

The degree of order in an alloy may be defined in a number of different ways; the two generally used parameters are those due to Bragg and Williams and to Bethe. Confining themselves to crystals of the type shown in Fig. 5.1, wherein there are only two sublattices, Bragg and Williams defined the degree of long-range order S as

$$S = \frac{A_\alpha - A_\alpha{}^r}{A_\alpha{}^\circ - A_\alpha{}^r} = \frac{B_\beta - B_\beta{}^r}{B_\beta{}^\circ - B_\beta{}^r} \tag{5.1}$$

Here A_α is the number of A atoms on α sublattice sites in a given solution, $A_\alpha{}^r$ is the number which would be on α sites if the solution contained no long-range order, $A_\alpha{}^\circ$ the number if the solution were perfectly ordered, and B_β, $B_\beta{}^r$, and $B_\beta{}^\circ$ have the same significance for B atoms and the β sublattice. S may be seen to vary from zero for no long-range order $(A_\alpha = A_\alpha{}^r)$ to unity for perfect order $(A_\alpha = A_\alpha{}^\circ)$.

The B-W order parameter is called the long-range order parameter because it concerns itself explicitly only with the extent to which the various sublattices are properly occupied over long distances. For any given value of S other than 1, however (and even for $S = 1$ in alloys where the composition is not at the ideal 50:50 ratio), there may be distinguished many different ways in which the A and B atoms on each sublattice may be arranged. For example, the A and B atoms may be randomly arranged within each sublattice, or they may have various degrees of local order within each sublattice. Any of these local arrangements could be consistent with a single intermediate value of S, but each would represent different degrees of short-range order. Recognizing this, Bethe introduced as a convenient measure of the local order a short-range order parameter σ, which he defined as

$$\sigma = \frac{n - n^r}{n^\circ - n^r} \tag{5.2}$$

Here n is the number of unlike nearest-neighbor atom pairs in a given state of a solid solution, n^r the number which would be present if the solution were completely random, and n° the number if the solution were perfectly ordered. σ, like S, varies from zero to unity; it is zero when the solution is completely random $(n = n^r)$

and unity when the solution is prefectly ordered ($n = n°$). However, σ and $\mathcal{S}$ have the same numerical values only in the completely ordered solutions. In solutions of intermediate order the exact relationship between the two depends on the particular assumptions made with regard to the local arrangements on each long-range sublattice, as will be seen later. A striking case in point is that in Fig. 5.1, where it may be seen that $\mathcal{S}$ is zero but σ is 0.89!

Before proceeding to the discussion of the mathematical relationship between the equilibrium order parameters and temperature, it seems worthwhile to digress in order to clarify some aspects of the nomenclature attached to ordering about which some confusion seems to exist. In particular, the terms "Bragg-Williams theory of ordering" and "Bethe theory of ordering" have each been rather vaguely and interchangeably applied to the three different concepts: (1) *physical model* of ordering, (2) *order parameter*, and (3) *mathematical treatment* of the degree of order versus temperature. The B-W *model* of ordering led to the B-W *long-range order parameter* $\mathcal{S}$. Since this parameter can be, and is, used to describe the degree of long-range order in other models as well, to avoid confusion it should properly be called simply the long-range order parameter; we shall adopt this practice here. The B-W *mathematical treatment* for the equilibrium degree of order versus temperature is based on the B-W model, the long-range order parameter $\mathcal{S}$, and the implicit simplifying assumption that the A and B atoms, though tending to separate onto the appropriate sublattices, are randomly distributed within each sublattice. Similarly, the Bethe *model* of ordering led to the Bethe *short-range order parameter* σ, which we shall henceforth call simply the short-range order parameter. The Bethe *mathematical treatment* for the equilbrium degree of order versus temperature, along with the Q-C treatment, is based on the tendency to form unlike nearest-neighbor pairs and leads to a specific relationship between σ and $\mathcal{S}$ and to equations for both σ and $\mathcal{S}$. The $\mathcal{S}$ obtained in this case does not, however, have, the same value at a given temperature as the $\mathcal{S}$ derived by the B-W treatment, since the assumptions in the two treatments with regard to local order are quite different. A relation between σ and $\mathcal{S}$ for the B-W assumption can also be derived, although Bragg and Williams themselves did not do this.

The appropriate interrelationship between the various B-W, Bethe, and Q-C terms discussed above may be summarized as follows:

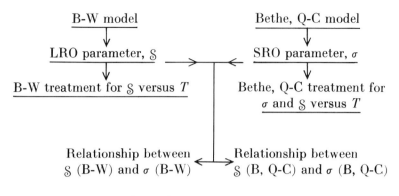

$$\text{At a given temperature}$$
$$\text{S (B-W)} \neq \text{S (B, Q-C)}$$
$$\sigma \text{ (B-W)} \neq \sigma \text{ (B, Q-C)}$$

5.4 QUASI-CHEMICAL CALCULATION OF THE EQUILIBRIUM DEGREE OF ORDER VERSUS TEMPERATURE

We shall confine our detailed attention to the general composition for the bcc-based structure shown in Fig. 5.1. Even for this simple case the problem has been treated only approximately; for other structures the mathematics becomes prohibitively tedious, and the simplifying assumptions necessary cast considerable doubt on the validity of the results. The notations to be used will be, as before,

N = total no. of lattice sites = total no. of atoms

$\dfrac{N}{2}$ = no. of α sites = no. of β sites

A_α = no. of A atoms on α sites

A_β = no. of A atoms on β sites

B_β = no. of B atoms on β sites

B_α = no. of B atoms on α sites

n_{AA} = no. of AA pairs

n_{BB} = no. of BB pairs

n_{AB} = no. of AB pairs, A on α and B on β

n_{BA} = no. of AB pairs, A on β and B on α

$n = n_{AB} + n_{BA}$

We wish to obtain the Q-C expression for the free energy of the solid solution as a function of composition, long-range order $\S$, short-range order σ, and temperature T. Then, by minimizing the free energy with respect to both $\S$ and σ, we shall find two equations the simultaneous solution of which will give the equilbrium values of $\S$ and σ at fixed composition and temperature. Solution at different temperatures then will allow the establishment of the equilibrium degree of order–temperature curves, *as predicted by the Q-C model.*

As in Eq. (4.38), the free energy of the solution is

$$G^S = G^M + \Delta G^{xs} - T \, \Delta S_m \tag{4.38}$$

To find ΔG^{xs}, we again make the assumptions of regular solution and ordinary pressures, so that for condensed phases

$$\Delta G^{xs} = \Delta H^{xs} = E^{PE,S} - E^{PE,M}$$

as in Eq. (4.41). In terms of the numbers of nearest-neighbor pairs

$$\Delta H^{xs} = (n_{AA} - n_{AA}{}^M)\mho_{AA} + (n_{BB} - n_{BB}{}^M)\mho_{BB} \\ + (n_{AB} + n_{BA})\mho_{AB} \tag{5.3}$$

where the superscript M refers to the mechanical mixture of the pure components and the terms without superscript refer to the solution. Since the pure components do not involve ordering, $n_{AA}{}^M$ and $n_{BB}{}^M$ are still given by the values in Eq. (4.46)

$$n_{AA}{}^M = \frac{ZN}{2} X_A \mho_{AA} \tag{5.4}$$

$$n_{BB}{}^M = \frac{ZN}{2} X_B \mho_{BB} \tag{5.5}$$

The numbers of the various kinds of nearest-neighbor pairs in the solution do depend on the short-range order and must therefore be expressed in terms of σ. To simplify this problem let us first find a relationship between n_{AA} and n_{BB} on the one hand and n on the other. The total number of pairs made by A atoms on α sites is ZA_α and by A atoms on β sites is ZA_β. These are plainly

$$ZA_\alpha = n_{AA} + n_{AB} \tag{5.6}$$

and

$$ZA_\beta = n_{AA} + n_{BA} \tag{5.7}$$

Adding, noting that $A_\alpha + A_\beta = NX_A$ and $n_{AB} + n_{BA} = n$, and rearranging gives

$$n_{AA} = \tfrac{1}{2}(ZNX_A - n) \tag{5.8}$$

Similarly,

$$ZB_\alpha = n_{BB} + n_{BA} \tag{5.9}$$

and

$$ZB_\beta = n_{BB} + n_{BA} \tag{5.10}$$

from which

$$n_{BB} = \tfrac{1}{2}(ZNX_B - n) \tag{5.11}$$

Upon substitution of (5.4), (5.5), (5.8), and (5.11) into (5.3) we have

$$\Delta H^{xs} = n\upsilon \tag{5.12}$$

where, as before,

$$\upsilon = \upsilon_{AB} - \frac{\upsilon_{AA} + \upsilon_{BB}}{2}$$

The value of n in terms of σ can be found easily by solving Eq. (5.2) for n,

$$n = (n^\circ - n^r)\sigma + n^r$$

Since in the completely ordered solution every A atom is completely surrounded by B atoms ($X_A \leqq X_B$),

$$n^\circ = ZNX_A \tag{5.13}$$

and, from Eq. (4.42),

$$n^r = ZNX_AX_B \tag{5.14}$$

Thus

$$n = ZNX_A(\sigma X_A + X_B) \tag{5.15}$$

and, substituting (5.15) into (5.12),

$$\Delta H^{xs} = ZNX_A\upsilon(\sigma X_A + X_B) \tag{5.16}$$

The similarity between Eqs. (5.16) and (4.47) is apparent; Eq. (5.16) reduces to Eq. (4.47) for $\sigma = 0$.

It may be seen from Eq. (5.16) that ΔH^{xs}, which represents the difference in internal energy between the solution and the mechanical mixture of the pure components, depends only on the degree of short-range order σ and not on the long-range order $\mathcal{S}$. This is to be expected in the Q-C view, since in this view the internal energy is made up only of nearest-neighbor interaction energies, and σ, not $\mathcal{S}$, expresses the number of the various types of nearest-neighbor pairs. On the other hand, the mixing entropy ΔS_m, the *configurational* entropy, does depend on $\mathcal{S}$ as well as on σ. This entropy is given by the Boltzmann equation as

$$\Delta S_m = k \ln W \tag{5.17}$$

where W is the number of different ways the atoms may be arranged to give a specific macroscopic state of the system. It is possible to vary σ and $\mathcal{S}$ independently; i.e., it is possible to distinguish between arrangements having a given $\mathcal{S}$ and different values of σ and also between arrangements having a given σ and different values of $\mathcal{S}$. Thus, both must be specified in calculating W for a given state. W is, then, the number of ways the $X_A N/2$ A atoms and the $X_B N/2$ B atoms may be simultaneously placed on the $N/2$ α sites and the $N/2$ β sites to give arrangements with a specified $\mathcal{S}$ and σ. This calculation has been carried out only approximately, typically by assuming that W is proportional to the number of ways the nearest-neighbor pairs can be divided among the four sets AA,

BB, AB, and BA to give the specified $\mathcal{S}$ and σ. Statistical theory gives for this

$$W = CW^{\text{pairs}} = C\,\frac{(ZN/2)!}{n_{\text{AA}}!\,n_{\text{AB}}!\,n_{\text{BA}}!\,n_{\text{BB}}!} \tag{5.18}$$

where C is a proportionality constant. To determine C, it is noted that the exact value of W for one particular state can be calculated independently of Eq. (5.18). This is the state in which for any specified $\mathcal{S}$ the atoms are randomly distributed on each sublattice (the state implicitly assumed by Bragg and Williams). In this case

$$W = W^{\text{atoms}} = \frac{(N/2)!}{A_\alpha!\,B_\alpha!}\,\frac{(N/2)!}{A_\beta!\,B_\beta!} \tag{5.18a}$$

It is, thus, assumed that C should have the value which produces (5.18a) from (5.18) for this random case. In such a state W has its maximum value for a given $\mathcal{S}$; therefore, differentiating (5.18) with respect to σ and maximizing,

$$\frac{\partial \ln W}{\partial \sigma} = 0 \qquad \mathcal{S}\ \text{const}$$

from which

$$\frac{(X_{\text{A}}\sigma + X_{\text{B}} + \mathcal{S})(X_{\text{A}}\sigma + X_{\text{B}} - \mathcal{S})}{(1 - \sigma)(X_{\text{B}}{}^2 - X_{\text{A}}{}^2\sigma)} = 1$$

The solution of this for all compositions is

$$\sigma = \mathcal{S}^2 \tag{5.18b}$$

(which, it should be noted, is the relationship between σ and $\mathcal{S}$ for the B-W condition). Substituting (5.18b) into (5.18) and equating to (5.18a), it is found that

$$C = (W^{\text{pairs}})^{1-Z} \tag{5.18c}$$

The appropriate expression for ln W in terms of σ and $\mathcal{S}$ is then found from (5.18c) and (5.18) as

$$
\begin{aligned}
\ln W = -\frac{ZN}{2}\{ & X_A^2(1-\sigma)\ln[X_A^2(1-\sigma)] \\
& + X_A(\S + X_A\sigma + X_B)\ln[X_A(\S + X_A\sigma + X_B)] \\
& + X_A(-\S + X_A\sigma + X_B)\ln[X_A(-\S + X_A\sigma + X_B)] \\
& + (X_B^2 - X_A^2\sigma)\ln(X_B^2 - X_A^2\sigma)\} \\
+ \frac{(Z-1)N}{2}\{ & X_A(1+\S)\ln[X_A(1+\S)] \\
& + X_A(1-\S)\ln[X_A(1-\S)] \\
& + [1 - X_A(1+\S)]\ln[1 - X_A(1+\S)] \\
& + [1 - X_A(1-\S)]\ln[1 - X_A(1-\S)]\} \quad \textbf{(5.19)}
\end{aligned}
$$

The final Q-C expression for the free energy of the solution is obtained by combining Eqs. (4.38), (4.41), (5.16), (5.17), and (5.19), yielding

$$
G^S = G^M + ZNX_A\upsilon(\sigma X_A + X_B) - kT\ln W \quad \textbf{(5.20)}
$$

where $\ln W$ is given by (5.19). To find the equilibrium values of σ and $\S$ as a function of T for a given composition, G^S is minimized with respect to both σ and $\S$ by setting

$$
\frac{\partial G}{\partial \sigma} = 0 \qquad \S, T \text{ const} \quad \textbf{(5.21)}
$$

and

$$
\frac{\partial G}{\partial \S} = 0 \qquad \sigma, T \text{ const} \quad \textbf{(5.22)}
$$

The resulting equations are

$$
-\frac{kT}{2}\ln\frac{(X_A\sigma + X_B) - \S^2}{(1-\sigma)(X_B^2 - \sigma X_A^2)} = \upsilon \quad \textbf{(5.23)}
$$

and

$$(Z - 1) \ln \frac{(1 + \mathcal{S})[1 - X_A(1 - \mathcal{S})]}{(1 - \mathcal{S})[1 - X_A(1 + \mathcal{S})]}$$

$$= Z \ln \frac{X_A\sigma + X_B + \mathcal{S}}{X_A\sigma + X_B - \mathcal{S}} \quad (5.24)$$

From the conditions that $\mathcal{S} = 0$ at $T \geq T_c$ and $\mathcal{S} \neq 0$ at $T < T_c$, a value of υ in terms of T_c can be found. This is

$$\upsilon = -\frac{kT_c}{2} \ln \frac{Z^2 X_A X_B}{(ZX_A - 1)(ZX_B - 1)} \quad (5.25)$$

Substitution of (5.25) into (5.23) gives

$$\frac{T}{T_c} = \frac{\ln \dfrac{Z^2 X_A X_B}{(ZX_A - 1)(ZX_B - 1)}}{\ln \dfrac{(X_A\sigma + X_B)^2 - \mathcal{S}^2}{(1 - \sigma)(X_B{}^2 - \sigma X_A{}^2)}} \quad (5.26)$$

Simultaneous solution of Eqs. (5.24) and (5.26) at a given composition yields the equilibrium curves of both $\mathcal{S}$ and σ versus the reduced temperature T/T_c.

Plots of these curves for the 50:50 alloy shown in Fig. 5.1 ($X_A = X_B = \frac{1}{2}$, $Z = 8$) are exhibited by the full lines in Fig. 5.4. It may be seen that at low temperatures the order parameters decrease slowly with rise in temperature; at higher temperatures they decrease more rapidly. In other words, the rate of change of the order parameter with temperature is an inverse function of the parameter, a characteristic of *cooperative* phenomena. The reason for this may be understood by reference to Eq. (5.16), which gives the excess enthalpy for the solution with order σ. ΔH^{xs} is the enthalpy contribution toward the lowering of the free energy which is attendant upon changing the alloy from the state in which it contains only AA and BB pairs (the mixture of pure components) to that containing $ZNX_A(\sigma X_A + X_B)$ AB pairs. In other words, ΔH^{xs} is a measure of the *strength* of the tendency toward ordering, and this is seen from Eq. (5.16) to be a function of the degree of order. Thus, as the degree of order decreases with rising temperature,

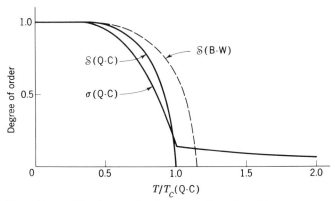

Fig. 5.4   *The degree of order versus the reduced temperature T/T_c (Q-C) for a 50:50 alloy based on a bcc disordered structure. The curves for long-range order according to the Bragg-Williams treatment, S (B-W), and for long-range order, S (Q-C), and short-range order, σ (Q-C), according to the quasi-chemical treatment are shown.*

the strength of the ordering tendency also decreases. This makes it correspondingly easier for the next increment of disordering to take place and finally leads to a catastrophic decrease in the degree of order as the critical temperature is approached. In terms of nearest-neighbor bonds we may say that at low temperatures, where nearly all bonds are AB bonds, i.e., the strongest bonds, placing one atom in a wrong position means breaking a maximum number of these strongest bonds and replacing them with weaker AA and BB bonds. At higher temperatures the number of AB bonds per atom is smaller, since σ is smaller. Thus, to move an atom from a right to a wrong position replaces fewer AB bonds with AA or BB bonds. Each of these processes requires an energy (strictly enthalpy) input, but in the higher temperature case the amount required is smaller. Hence the process is easier: a given increase in thermal energy (rise in temperature) produces a greater number of interchanges; accordingly, σ and S decrease faster as the temperature increases.

The order-disorder transition with the theoretical characteristics indicated in Fig. 5.4 is not a first-degree transformation.* Thus, it should probably not be viewed as a transformation from one phase to another but rather as a continuous change in state of

* See pages 126–129 for ordering with first-degree characteristics.

a given phase as the temperature is changed. It is instructive, nevertheless, to consider the transition in a manner analogous to that illustrated for first-degree transformations by the free-energy–temperature curves in Fig. 3.6. To do this let us suppose that each of the infinite series of states through which, say, a disordering phase passes as the temperature is raised is a separate "phase," characterized by a given degree of order, which is the same at all

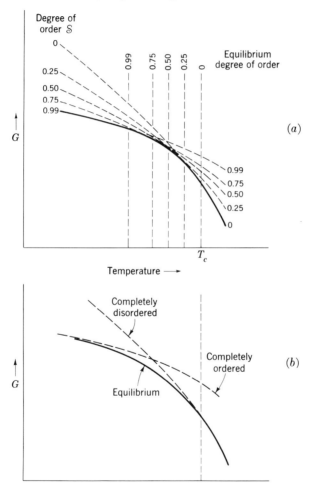

Fig. 5.5 *Schematic curves of free energy versus temperature for a series of hypothetical alloys having various degrees of long-range order (dashed lines), and the resulting lowest-free-energy curve (full line).*

temperatures, and by its own G-T curve. There would then be an infinite number of G-T curves each differing only infinitesimally in slope from its neighbors. The equations for these curves would be given, in the treatment outlined above, by Eq. (5.20), each with a different pair of values of σ and $\mathcal{S}$. A group of such curves is illustrated in Fig. 5.5a. It may be seen that as the temperature is raised, the material will transform in a series of infinitesimal steps from one "phase" to the next as the free-energy curve of each successive "phase" becomes the lowest of all the "phases." The full line in Fig. 5.5a gives the locus of the lowest free energy as a function of temperature. The relationship between the G-T curves of the completely ordered and the completely disordered "phases" is shown in Fig. 5.5b. It should be noted that though the point of intersection of these two curves indicates the temperature at which a completely ordered and a completely disordered "phase" would be in equilibrium, this is not the critical temperature and in fact represents only an unstable equilibrium, for even the smallest ordering of the disordered "phase" or the smallest disordering of the ordered "phase" tends to lower the free energy.

5.5   THE BRAGG-WILLIAMS APPROXIMATION

In their early treatment of the degree of order versus temperature Bragg and Williams assumed, as pointed out on page 119, that the A and B atoms separated onto their appropriate sublattices but were arranged at random within the sublattices. Thus, the B-W $\mathcal{S}$-T curve can be obtained from Eqs. (5.19) and (5.20) by finding the value of σ which for a given $\mathcal{S}$ maximizes W (though Bragg and Williams themselves used a different method). From Eq. (5.19), setting

$$\left(\frac{\partial W}{\partial \sigma}\right)_{\mathcal{S}} = 0$$

it may be found that this condition yields Eq. (5.18b)

$$\sigma = \mathcal{S}^2 \tag{5.18b}$$

This is the relationship which Bragg and Williams implicitly assumed for σ and $\mathcal{S}$. Substitution of Eq. (5.18b) into (5.20) and minimization of G with respect to $\mathcal{S}$ gives the B-W $\mathcal{S}$-T equation as

$$-\frac{Z\upsilon}{kT}\,4X_{\mathrm{A}}\mathcal{S} = \ln\frac{(1+\mathcal{S})[1-X_{\mathrm{A}}(1-\mathcal{S})]}{(1-\mathcal{S})[1-X_{\mathrm{A}}(1+\mathcal{S})]} \tag{5.27}$$

Again, using the condition $\mathcal{S} = 0$ at $T \geqq T_c$, it is found that

$$\upsilon = -\frac{kT_c}{2X_{\mathrm{A}}X_{\mathrm{B}}Z} \tag{5.28}$$

so that combining (5.27) and (5.28) gives

$$\frac{T}{T_c} = \frac{2\mathcal{S}}{X_{\mathrm{B}}\ln\dfrac{(1+\mathcal{S})[1-X_{\mathrm{A}}(1-\mathcal{S})]}{(1-\mathcal{S})[1-X_{\mathrm{A}}(1+\mathcal{S})]}} \tag{5.29}$$

This is the B-W prediction of $\mathcal{S}$ versus the reduced temperature T/T_c. For comparison with the more realistic Q-C (Bethe) prediction, Eq. (5.29) is plotted in Fig. 5.4 for the 50:50 (bcc-based) alloy. There are two major differences between the two curves for $\mathcal{S}$ versus T. The first is seen to lie in the higher T_c predicted by the B-W approximation; from Eqs. (5.25) and (5.28) the relationship is

$$T_c\,(\text{B-W}) = 1.14T_c\,(\text{Q-C}) \tag{5.30}$$

More significantly, the B-W treatment does not account at all for the existence of short-range order above T_c; this failing, of course, follows directly from the fact that the B-W treatment was based not on nearest-neighbor pair (short-range) interactions but rather on some unspecified long-range atomic forces.

5.6   ORDERING AND THE PHASE DIAGRAM

Equations (5.25) and (5.28) allow the calculation of the Q-C and B-W theoretical curves for the critical temperature versus composition for the bcc-based structure. In terms of the critical temperature for the 50:50 alloy, Eq. (5.25) gives (for $Z = 8$)

$$\frac{T_c}{T_c\,(X=0.5,\ \text{Q-C})} = 0.574\ln\frac{64X_{\mathrm{A}}X_{\mathrm{B}}}{(8X_{\mathrm{A}}-1)(8X_{\mathrm{B}}-1)} \tag{5.31}$$

Similarly, Eq. (5.28) gives

$$\frac{T_c}{T_c\,(X\,=\,0.5,\ \text{B-W})} = 4X_A X_B \tag{5.32}$$

which, in view of Eq. (5.30), becomes

$$\frac{T_c\,(\text{B-W})}{T_c\,(X\,=\,0.5,\ \text{Q-C})} = 4.56 X_A X_B \tag{5.33}$$

When Eqs. (5.31) and (5.33) are plotted, as in Fig. 5.6, they produce symmetrical curves very much like the miscibility gap discussed in Chap. 4. In this case, however, the critical-temperature curve does not enclose a region of two phases but rather a region in which the properties of a single phase change with temperature. The line in the diagram simply marks the locus of temperatures at which long-range order finally is gone on heating or just starts to appear on cooling.

For alloys with structures other than the simple bcc-based structure it is not possible to carry out even approximately correct analyses on the relatively simple basis described above without making very questionable assumptions. More sophisticated, but still

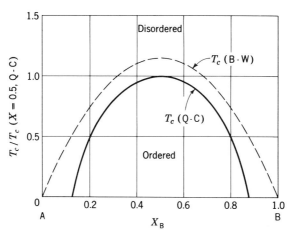

Fig. 5.6   *Temperature-composition diagram for a bcc-based alloy showing the long-range-order critical temperature according to the Bragg-Williams treatment and the quasi-chemical treatment.*

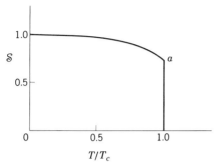

Fig. 5.7   *Schematic curve of long-range order versus temperature for an fcc-based structure.*

approximate, treatments, the mathematics of which is beyond the scope of this book, have been devised for these structures. These treatments predict that for other structures the long-range order transition has both first-degree and higher-degree characteristics. For example, in the simple fcc-based structure it is predicted that the S-T curve looks as shown schematically in Fig. 5.7; the degree of long-range order decreases gradually at first as the temperature is raised but then drops discontinuously at T_c from some finite value, such as a in the figure, to zero. Since the curve in Fig. 5.7 is an equilibrium curve, it is apparent that at T_c two phases, one with $S = a$ and one with $S = 0$, may be in equilibrium with each other. Thus, the two phases can coexist at T_c with a distinct physical boundary between them, a first-degree characteristic. This results in the appearance of two-phase fields in the phase diagram for this type of order phenomenon, as illustrated in Fig. 5.8. It is also noteworthy that the theory predicts for the fcc-based structure the

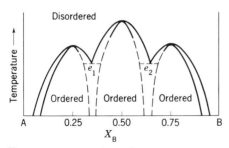

Fig. 5.8   *Possible phase diagram for an fcc-based alloy system where three ordered structures occur centered on the 25, 50, and 75 atomic percent alloys.*

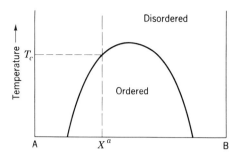

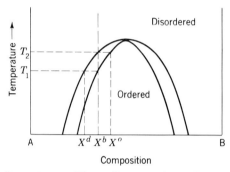

Composition

Fig. 5.9   *Phase diagrams for ordering systems in which two-phase stability occurs (bottom) and does not occur (top) during equilibrium heating or cooling.*

possible occurrence of three ordered phases centering on the three compositions A_3B, AB, and AB_3, as shown in the figure. This, of course, introduces the likelihood of invariant reactions between the ordered phases, as indicated by the dashed lines in Fig. 5.8.

The difference between the transitions undergone near the critical temperature by alloys of the system in Fig. 5.8 and those in Fig. 5.6 may be illustrated by following, say, the heating of an ordered alloy in each. Referring to Fig. 5.9, on heating the alloy of composition X^a the degree of long-range order decreases monotonically, as shown in Fig. 5.4, until at T_c for this alloy it has become zero. There is no constant-temperature change in enthalpy, i.e., no latent heat, associated with this transition but rather an unusually rapid increase in the heat capacity (absorption of energy by the disordering process) as the disordering takes place. For the alloy of composition X^b in Fig. 5.9, the same type of phenomenon takes place up to the temperature T_1, which is T_c for the composition

X^b, but at T_1 the ordered phase, characterized by long-range order $\mathcal{S} = a$, becomes saturated with respect to the disordered phase with no long-range order and of composition X^d. The latter, therefore, forms by a process of nucleation and growth just as a second phase does in traversing other two-phase regions. As heating is continued, the temperature rises, and the ordered and disordered phases both shift composition along the appropriate boundaries of the two-phase region, again just as in other two-phase, two-component phase changes. The transition is finally over when the temperature has reached T_2, where only the newly formed phase with long-range order equal to zero remains. The composition of the disordered phase will now be X^b, and that of the last ordered phase to disappear will have been X^a. (The "disordered" phase, of course, will still exhibit some short-range order, so that it is not correct to say it is completely disordered.) It is worth emphasizing that the ordered phase starts its first-degree transition at T_c, and though the temperature continues to rise during this transition, the ordered phase continues to be at T_c; this is because as the composition of the ordered phase changes, so does its critical temperature and in such a way that the temperature of the phase is always equal to its critical temperature.

Little has been said in this chapter about the agreement between experiment and the theoretical curves discussed here. A comparison of Figs. 4.33 to 4.35 with Figs. 5.6 and 5.8 reveals that there is good qualitative agreement; quantitative data are, however, quite sparse, but those available show only fair agreement with even the more sophisticated theories. For reviews of this subject, Refs. 5.6, 5.7, and 5.8 are recommended.

Problems

5.1 How many atoms per lattice point are there in the *disordered* bcc structure corresponding to Fig. 5.1? How many in the ordered structure? What is the ratio of the unit cell-edge lengths for the disordered versus ordered structures?

5.2 (a) How many second-nearest neighbors does each atom have in a bcc structure (assuming there is, as in many metals, one atom per lattice point)? Third-nearest neighbors? (b) What are the first-, second-, and third-nearest-neighbor distances in the structure in terms of the unit cell-edge length a?

5.3 On the simple assumptions that the interaction energies between atom pairs vary inversely as the sixth power of their separations and that the total interaction energy of an array is the sum of the first-, second-, and third-nearest-neighbor interactions, calculate the percent of the total energy accounted for by first-, second-, and third-nearest neighbors for the structure of Prob. 5.2. Compare with the statement in parentheses on page 80, and comment.

5.4 Show that Eq. (5.18*b*) is the solution of the equation just above it.

5.5 Show that ΔS_m as given by Eqs. (5.17) and (5.19) reduces to Eq. (4.3) for both $\mathcal{S}$ and $\sigma = 0$.

5.6 Substitute $\sigma = \mathcal{S}^2$ into Eq. (5.20) to find the B-W expression for G^s versus T and X. *Hint:* Substitution of $\sigma = \mathcal{S}^2$ into Eq. (5.19) should cause the terms in $ZN/2$ to cancel, thus leaving only the $-N/2$ terms as the B-W expression for ln W.

5.7 Using Eq. (5.28) and the expression developed in Prob. 5.6, sketch curves of $G^s - G^M$ versus $\mathcal{S}$ (for the range $\mathcal{S} = -1$ to $+1$) for the 50:50 alloy at $T = 0$, $T = T_c/2$, $T = T_c$, and $T = \frac{5}{4}T_c$. Comment on the meaning of the shapes of these curves with respect to the equilibrium degree of order.

5.8 Using Eq. (5.25) and the data in Fig. 4.35, check the assumption that $\mathcal{U}$ is not a function of X in the range 30 to 90 atomic percent platinum in the Cu–Pt system.

References

5.1 C. S. Barrett and T. B. Massalski, "Structure of Metals," 3d ed., p. 271, McGraw-Hill Book Company, New York, 1966.

5.2 H. A. Bethe, *Proc. Roy. Soc. (London)*, **150A**:552 (1935).

5.3 J. C. Slater, *Phys. Rev.*, **84**:179 (1951).

5.4 J. F. Nicholas, *Proc. Phys. Soc. (London)*, **A66**:201 (1953).

5.5 W. L. Bragg and E. J. Williams, *Proc. Roy. Soc. (London)*, **145A**:699 (1934).

5.6 L. Guttman, Order-Disorder Phenomena in Metals, *Solid State Phys.*, **3**:146 (1956).

5.7 S. Seigel, Order-Disorder Transitions in Metal Alloys, in "Phase Transformations in Solids," John Wiley & Sons, Inc., New York, 1951.

5.8 T. Muto and Y. Takagi, The Theory of Order-Disorder Transitions in Alloys, *Solid State Phys.*, **1**:194 (1955).

SIX

TWO-COMPONENT SYSTEMS CONTAINING INVARIANT REACTIONS: THE EUTECTIC AND EUTECTIC-LIKE SYSTEMS

6.1   INTERMEDIATE PHASES

The discussion in Chap. 4 was limited to isomorphic alloy systems in which the only stable phases were solutions of the two components. This limitation requires not only that the two components have the same crystal structure, but also that they be chemically and physically very much alike. In practice, however, the components are most frequently of different crystal structure and significantly dissimilar in electrochemical behavior and/or atomic size; this generally leads to the appearance of stable intermediate phases and the limiting of the intersolubility of the two components.

When the electrochemical dissimilarity between the components is large, the (constant-pressure) free-energy curves may be as shown in Fig. 6.1 for three typical solid phases in a hypothetical system at some specific temperature. Here it has been assumed that the terminal phases α and γ are solid solutions with different crystal structures. In this case it should be noted that the α and γ curves intersect each other rather than joining smoothly in the central portion of the diagram as do the free-energy curves for isomorphous phases. In the isomorphous cases, for ΔH^{xs} negative, a single solid solution exists at all temperatures, and for ΔH^{xs} positive two solid solutions coexist at low temperatures, becoming one at high temperatures. When the crystal structures of the two components are different, however, no such possibility exists: at all temperatures where the solids may exist there must be some point between the compositions $X = 1$ and $X = 0$ where the structure changes from that of one component to that of the other; i.e., there must be a phase change. The free-energy curves appear, then, as in Fig. 6.1. In the discussions of the various phase-diagram forms in the remainder of this book, we shall adopt the procedure of representing terminal phases by intersecting free-energy curves such as those in Fig. 6.1. It should be kept in mind, however, that in systems of positive ΔH^{xs} from a qualitative point of view smoothly inflected curves such as those in Fig. 4.19 could equally well be used.

The phase β represented in Fig. 6.1 tends to have the nature of a chemical compound. Because of strong attractive forces between unlike atoms, the atoms in such phases arrange themselves in highly ordered patterns. In metal-metal and metal-nonmetal systems, these phases differ from ordered solutions in that the atomic bonding forces have strong ionic or covalent characteristics; as a rough rule of thumb, the greater the dissimilarity between the components, the less metallic and the more nonmetallic the bonding. As a result of the directed nature of these nonmetallic forces and the high order characteristic of the atomic arrays, the entropies of these intermediate phases are low, but the enthalpies are also very low,

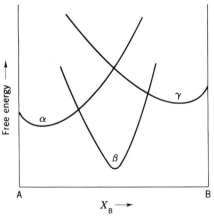

Fig. 6.1 *Typical free-energy curves for terminal phases and an intermediate phase in a system in which the electrochemical dissimilarity is large.*

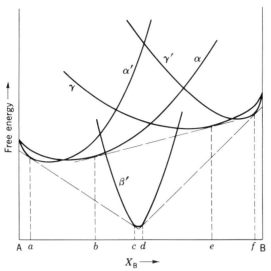

Fig. 6.2 *Restriction of the intersolubility of two components by the formation of a compoundlike intermediate phase.*

i.e., of high negative magnitude; the latter effect predominates and the phases, thus, have very low free energies, i.e., are highly stable. Because the greatest numbers of unlike atomic bonds can usually be obtained in crystal arrays within which the atomic fractions are proportional to small integers, and because the enthalpy per atomic bond is a strong function of composition in these phases, intermediate phases of this type tend to exist at compositions corresponding to simple stoichiometric ratios such as AB_2, AB, A_3B_2, etc. For the same reasons, the free energies rise rapidly as the composition varies from these ratios, giving curves with rather sharply marked minima, as illustrated for the β phase in Fig. 6.1.

The way in which the tendency to form compoundlike intermediate phases limits the intersolubility of components is illustrated in Fig. 6.2. In this figure it is supposed that there are, on the one hand, two components, A and B, with different crystal structures and having at some specific temperature in the solid range the free-energy curves marked α and γ in the figure. If it is further supposed that these two components are moderately dissimilar electrochemically, that ΔH^{xs} is negative, and that no intermediate phase is stable at this temperature, then the intersolubility limits are given at b and e by the common tangent to the α and γ free-energy curves.

On the other hand, if it is supposed that the two components have a considerably greater electrochemical dissimilarity and that a stable, compoundlike intermediate phase does form, the free-energy curves of the two terminal phases might then be represented by those marked α' and γ' and that of the intermediate phase by β' in Fig. 6.2. The intersolubility limits are then given at a and f and are seen to be drastically restricted by the presence of the phase β'. The diagram also shows why the composition limits of intermediate phases with sharply marked free-energy minima are narrow; the points c and d give these composition limits for the phase β'.

Intermediate solid phases may appear in alloy systems even when the interatomic forces between like atoms are stronger than those between unlike atoms. This is particularly true when the crystal structures of the components are different. It may happen that at intermediate compositions a *third crystal structure* may have a lower free energy than either of the two parent structures and also lower than a mixture of solid solutions having the parent structures. This may be due to changes in electronic structures with composition, to a lowering of the free energy per interatomic bond on forming the new structures, or to a decrease in atomic size-difference strain energy in the new arrangement. Since these intermediate phases tend to be random solid solutions, their formation is also favored by the mixing entropy term in the free-energy equation. As is typical of solid solutions, their free energies do not vary rapidly with composition, and as a result their compositional ranges of existence tend to be broad, as illustrated by the points of tangency c and d in Fig. 6.3. Their formation also tends to limit the intersolubility of the components, as a comparison of points a and f with points b and e in Fig. 6.3 will show. The effect, however, is not so strong as it is for compoundlike intermediate phases, because the free energies of solid-solution-type intermediate phases are usually not greatly different from those of the terminal phases.

On the basis of the principles discussed in the previous chapters, free-energy curves such as those in Figs. 6.1 to 6.3 may be used to derive equilibrium diagrams of all the various types met in practice by making appropriate assumptions about the shapes of the curves and the relative rates at which the curves rise (or descend) as the temperature is lowered (or raised). The phase diagrams we shall discuss now differ from those already described primarily in that, with or without intermediate phases, they contain one or more *invariant regions*, i.e., regions where the number of phases coexisting

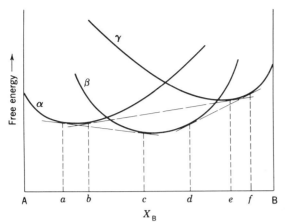

Fig. 6.3   *Typical free-energy curves in a system in which a solid-solution type of intermediate phase forms.*

in equilibrium is sufficient to reduce to zero the number of degrees of freedom. For two-component systems at arbitrarily fixed pressure, the number of phases required for invariance is, from the phase rule,

$$p = c + 1 - f = 2 + 1 - 0 = 3$$

The various possible types of invariant regions which may occur in two-component systems is not large, and the number actually found to occur is even smaller. They are usually classified according to two criteria: (1) whether the invariant reaction taking place on cooling through the region involves, on the one hand, a single phase decomposing to form two others or, on the other hand, two phases combining to form a third;* and (2) whether, and how many, liquids are involved in the reaction.† On the basis of the first criterion, the prototype, and by far the most commonly occurring, invariant reaction in one category is called the *eutectic* reaction. It involves on cooling the decomposition of a liquid phase into two solid phases or, perhaps more accurately, the separation of two solid phases from a liquid phase. The prototype in the other category is called the *peritectic* reaction and consists on cooling of the reaction be-

---

* As will be shown later, these reactions in two-component systems at arbitrarily fixed pressure in general involve either the appearance or the disappearance of one phase only (see boundary rule, page 185).

† We are, of course, considering only condensed phases; invariant reactions involving gas phases also exist.

tween a solid phase and a liquid phase to form a second solid phase. We shall discuss these two prototype reactions in considerable detail and follow with a much briefer treatment of the other additional types which are known to occur in each category (three eutectic-like and two peritectic-like reactions).

6.2   THE EUTECTIC SYSTEM

The eutectic invariant reaction may occur, and is most readily studied, in a system which contains only two terminal solid-solution phases and a single liquid phase. This invariant will occur in such a system whenever ΔG^{xs} is positive and $\Delta G^{xs,S}$ is enough larger than $\Delta G^{xs,L}$ to give rise to free-energy curves with the relationships and the temperature variations illustrated schematically in Fig. 6.4. The student may verify for himself that the free-energy curves shown lead to the equilibrium diagram in the lower portion of Fig. 6.4. In this diagram there are three regions of one-phase stability, three of two-phase stability, and one of three-phase stability. The one-phase and two-phase regions are similar to those discussed in Chap. 4. The three-phase region is one-dimensional, the line jkl at the temperature T_3. This line is simply the common tie line connecting the equilibrium compositions j, k, and l of the three coexisting phases. It arises from the fact that at T_3 the free-energy curves of the three phases α, liquid, and β are in such relative positions that a single line can be drawn tangent to all three. The liquid of composition k is called the *eutectic* liquid, from the Greek word *eutektos* meaning "easily fused"; it may be seen that liquid of composition k is the lowest-melting liquid in the alloy system and, in this sense, is most easily fused.

The close relationship between the eutectic phase diagram and the phase diagram previously examined in Fig. 4.21c, characterized by a liquidus minimum and a solid miscibility gap, is demonstrated in Fig. 6.5. In Fig. 6.5a, a diagram similar to that in Fig. 4.21c is shown. In Fig. 6.5b a qualitatively identical diagram is depicted, but it has been supposed that $\Delta G^{xs,S}$ is now enough larger than $\Delta G^{xs,L}$ so that the miscibility gap extends upward to temperatures where liquid is still stable; i.e., the miscibility gap and the liquid-solid region intersect. Let us now consider what happens to the intersecting lines in Fig. 6.5b in the light of the phase rule and the thermodynamic meanings of the lines involved. The portion of the

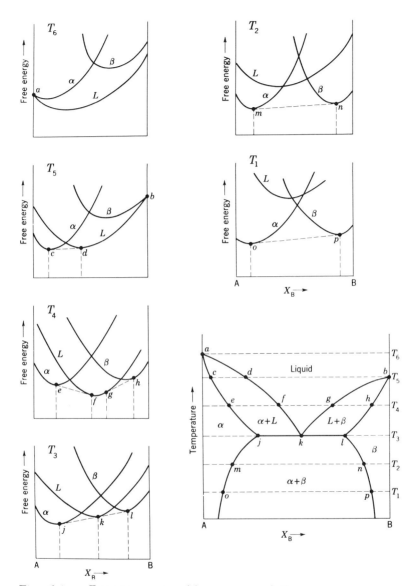

Fig. 6.4   *Free-energy–composition curves and the temperature-composition equilibrium diagram for a eutectic system.*

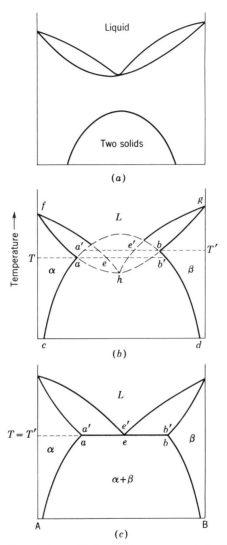

Fig. 6.5   *Similarity between a eutectic diagram and a diagram exhibiting a liquidus minimum and a solid solubility gap.*

solvus line ac represents compositions of α in equilibrium with β having compositions along the portion of the solvus line $b'd$. Thus, α of composition a, which we shall represent by the symbol α_a, is in equilibrium with $\beta_{b'}$, at the temperature T. The solidus line af represents compositions of α in equilibrium with liquid of compositions lying along the liquidus line fh; thus, α_a is in equilibrium with L_e at the temperature T. The temperatures and pressures of α_a, $\beta_{b'}$, and L_e are all identical; since the chemical potentials of phases in equilibrium are also identical and both $\beta_{b'}$ and L_e are in equilibrium with α_a, the chemical potentials of $\beta_{b'}$ and L_e must also be equal; that is, $\beta_{b'}$ and L_e are in equilibrium with each other. Therefore, all three phases, α_a, $\beta_{b'}$, and L_e, are in mutual equilibrium at the temperature T. Similarly β_b, $L_{e'}$, and $\alpha_{a'}$ are in mutual equilibrium at T'. But the phase rule tells us that in a two-component system at arbitrarily fixed pressure, the equilibrium between three phases is invariant. This means that for any three particular phases neither the temperature nor any of the phase compositions may at equilibrium be arbitrarily chosen, or, in other words, there is only one temperature and one composition for each of three phases at which they can coexist in equilibrium. Hence, the two temperatures T and T' must be the same, and the composition $a = a'$, the composition $b = b'$, and $e = e'$. When, in accordance with this conclusion, the lines in Fig. 6.5b are appropriately redrawn with the one addition that the tie line connecting the compositions a, e, and b is also drawn in, the eutectic diagram in Fig. 6.5c results.

6.3   CALCULATION OF THE MELTING-POINT CHANGE AND THE DISTRIBUTION COEFFICIENT IN DILUTE SOLUTIONS

We have shown in Chap. 4 that the liquidus and solidus curves in ideal systems can be calculated in a straightforward manner. In nonideal systems similar calculations become quite difficult because of the complexity involved in expressing ΔG^{xs} as a function of composition and temperature. It has been found empirically, however, that for sufficiently dilute solutions in all real systems the solvent behaves ideally (Raoult's law); i.e., the properties of the solvent atoms are largely unaffected by the presence of a few isolated solute atoms. With the help of this empirical law an approximate, simple mathematical relationship can be developed between composition and change in a phase-transformation temperature for dilute solutions.

Designating the solvent in dilute solution as component A and the solute as B, we may, according to Raoult's law, apply Eq. (4.36) to the solvent, giving

$$\ln \frac{1 - X_B{}^\alpha}{1 - X_B{}^\beta} = \frac{\Delta H_A{}^{tr}}{R} \frac{T_A{}^{tr} - T}{T_A{}^{tr}T} \tag{6.1}$$

where $\Delta H_A{}^{tr}$ is the latent heat (enthalpy) and $T_A{}^{tr}$ is the transformation temperature for the pure solvent in any phase transition α-β in the system. Making use of the approximation that for $0 < X \ll 1$, $\ln (1 - X) \cong -X$, noting that for small X_B we may also approximate $T_A{}^{tr}T$ as $(T_A{}^{tr})^2$, and writing $T - T_A{}^{tr}$ as ΔT, Eq. (6.1) becomes

$$\frac{X_B{}^\alpha}{X_B{}^\beta} = 1 + \frac{\Delta H_A{}^{tr}}{R(T_A{}^{tr})^2} \frac{\Delta T}{X_B{}^\beta} \tag{6.2}$$

In dilute solution the ratios $\Delta T/X_B{}^\beta$ and $X_B{}^\alpha/X_B{}^\beta$ are found to be essentially independent of temperature. $X_B{}^\alpha/X_B{}^\beta$ near the transformation temperature of the pure solvent is the distribution coefficient k, referred to previously in the discussion of zone refining in Chap. 4.

Equation (6.2) can be used to calculate k when the latent heat of transformation and the unit change of the solvent transition temperature for a solute $\Delta T/X_B{}^\beta$ are known, or alternately it can be used to estimate values of $\Delta H_A{}^{tr}$ or of $\Delta T/X_B{}^\beta$ when the other quantities are known. It can also be used to check the reasonableness of certain features of phase diagrams. For example, let us apply the equation to the Ag–Cu phase diagram, a typical eutectic diagram, shown in Fig. 6.6. In the determination of phase diagrams* it is usually much easier to determine a liquidus line such as line *ac* in Fig. 6.6 than a solidus line such as *ab*. It is reasonable to expect, then, that the line *ac* in the Ag–Cu diagram is substantially correct; we may, however, ask: "How likely is it that the solidus line *ab* has the convex-upward shape shown rather than, say, the opposite curvature or a straight-line form?" To answer this question we need to know whether the slope of the solidus line near the melting point of copper should be greater than, equal to, or less than the slope of a straight

* See, for example, Ref. 6.1.

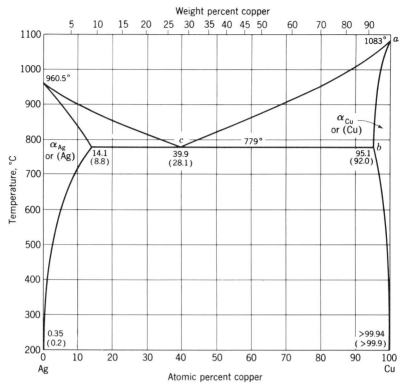

Fig. 6.6 *The Ag–Cu phase diagram. (From Hansen.[4.17])*

line drawn between *a* and *b*. The appropriate solidus slope is $\Delta T/X_{Ag}^{S}$; a value of this slope may be obtained by adapting Eq. (6.2) as follows:

$$\frac{\Delta T}{X_{Ag}^{L}} = \frac{\Delta T}{X_{Ag}^{S}}\left[1 + \frac{\Delta H_{Cu}^{m}}{R(T_{Cu}^{m})^{2}}\frac{\Delta T}{X_{Ag}^{L}}\right]$$

where the superscript *m* indicates values at the melting point. Inserting the values $T_{Cu}^{m} = 1356°K$ and $\Delta H_{Cu}^{m} = 3120$ cal/mole and, from the diagram, $\Delta T/X_{Ag}^{L} = -830$ corresponding to $-8.3°C$ per atomic per cent Ag near the melting point of pure copper, we find $\Delta T/X_{Ag}^{S}$ corresponds to $-28°C$ per atomic percent. The slope of a straight line from *a* to *b* would be $-61°C$ per atomic percent Ag, so that we see that the convex-upward curvature of line *ab* is corroborated by the thermodynamics of the system. Using the same data, k for silver in copper near the copper melting point is found to be 0.3.

It should be noted that the use of Eqs. (6.1) and (6.2) is not restricted to eutectic systems or only to systems in which a transition temperature is *lowered* by the addition of solute. They may be used more generally, the only restriction being that the solutions be sufficiently dilute for Raoult's law to apply. For example, they may be used in solid-solid transformations and in systems such as the peritectic, where, as will be seen, the melting point of one of the components is *raised* upon addition of solute. The range of dilution where Raoult's law may be expected to apply is not quantitatively predictable a priori; in general, this range is the more limited the more dissimilar the components.

6.4 CALCULATION OF THE SOLVUS LINE IN DILUTE SOLUTIONS

The solvus line in systems of limited terminal solubility can be described by the fairly simple equation

$$X_B{}^\alpha = \exp\left(-\frac{\Delta G^{\mathrm{xs},\alpha}}{RT}\right) = \exp\left(\frac{\Delta S^{\mathrm{xs},\alpha}}{R}\right)\exp\left(-\frac{\Delta H^{\mathrm{xs},\alpha}}{RT}\right) \quad (6.3)$$

where $X_B{}^\alpha$ is the equilibrium solubility of the solute B in the terminal solid solution α and $\Delta G^{\mathrm{xs},\alpha}$, $\Delta S^{\mathrm{xs},\alpha}$, and $\Delta H^{\mathrm{xs},\alpha}$ are the excess free energy, entropy, and enthalpy *per mole of B atoms* for B atoms in the α solution relative to B atoms in pure B.*

To derive Eq. (6.3), it may be recalled that the conditions of equilibrium between the two terminal phases are given by the common tangent to the appropriate free-energy curves. These conditions are expressed mathematically in Eqs. (4.13) and (4.23); restated they are

$$\frac{dG^\alpha}{dX_B} = \frac{dG^\beta}{dX_B} = \frac{G^\alpha - G^\beta}{X_B{}^\alpha - X_B{}^\beta} \quad (6.4)$$

---

* These excess quantities thus include the excess energy and entropy of transformation whenever the pure B has a different crystal structure than the α solid solution.

We may write for the free energies of the α and β terminal solid solutions

$$G^\alpha = X_A{}^\alpha \bar{G}_A{}^\alpha + X_B{}^\alpha \bar{G}_B{}^\alpha + RT(X_A{}^\alpha \ln X_A{}^\alpha + X_B{}^\alpha \ln X_B{}^\alpha) \tag{6.5}$$

$$G^\beta = X_A{}^\beta \bar{G}_A{}^\beta + X_B{}^\beta \bar{G}_B{}^\beta + RT(X_A{}^\beta \ln X_A{}^\beta + X_B{}^\beta \ln X_B{}^\beta) \tag{6.6}$$

from which

$$\frac{dG^\alpha}{dX_B} = \bar{G}_B{}^\alpha - \bar{G}_A{}^\alpha + RT \ln X_B{}^\alpha \tag{6.7}$$

and

$$\frac{dG^\beta}{dX_B} = \bar{G}_B{}^\beta - \bar{G}_A{}^\beta - RT \ln X_A{}^\beta \tag{6.8}$$

where use has been made of the approximation that, for α and β both dilute solutions, the individual free energies $\bar{G}_B$ and $\bar{G}_A$ in β and α are not functions of composition and

$$X_A{}^\alpha \cong 1 \cong X_B{}^\beta \qquad \text{and} \qquad X_A{}^\beta \ll 1 \gg X_B{}^\alpha$$

Similarly

$$\frac{G^\alpha - G^\beta}{X_B{}^\alpha - X_B{}^\beta} \cong G^\beta - G^\alpha = \bar{G}_B{}^\beta - \bar{G}_A{}^\alpha + X_A{}^\beta \bar{G}_A{}^\beta - X_B{}^\alpha \bar{G}_B{}^\alpha$$
$$+ RT(X_A{}^\beta \ln X_A{}^\beta - X_B{}^\alpha \ln X_B{}^\alpha) \tag{6.9}$$

Combining (6.7) and (6.9) gives

$$\bar{G}_B{}^\alpha - \bar{G}_B{}^\beta - X_B{}^\alpha \bar{G}_B{}^\alpha + X_A{}^\beta \bar{G}_A{}^\beta = RT(- \ln X_B{}^\alpha + X_A{}^\beta \ln X_A{}^\beta$$
$$- X_B{}^\alpha \ln X_B{}^\alpha) \tag{6.10}$$

In this equation, the first two terms on the left are large compared to the other two terms on the left, since the latter contain the small fractions $X_B{}^\alpha$ and $X_A{}^\beta$. The same is true of the first term on the right as compared with the sum of the other two terms. Thus, Eq. (6.10) simplifies to

$$\bar{G}_B{}^\alpha - \bar{G}_B{}^\beta = -RT \ln X_B{}^\alpha \qquad (6.11)$$

Since the β solid solution is dilute, the B atoms in it have essentially the same environment as B atoms in pure B. Thus, $\bar{G}_B{}^\beta$ is seen to be very nearly equal to G_B, and Eq. (6.11) is the equivalent of Eq. (6.3). By combining (6.8) with (6.9), the analogous equation for the solubility of A in β may be found:

$$\bar{G}_A{}^\beta - G_A = -RT \ln X_A{}^\beta$$

or

$$\Delta G^{xs,\beta} = -RT \ln X_A{}^\beta \qquad (6.12)$$

It has been pointed out[6.2] that Eqs. (6.3), (6.11), and (6.12) become quite inaccurate at solubilities exceeding about 1 to 2 atomic percent. Improvement on the approximations made in deriving these equations allowed the development of an analogous equation[6.2] for terminal solubilities

$$\frac{\ln\,[X_B{}^\alpha/(1 - X_A{}^\beta)]}{1 - 2X_B{}^\alpha} = -\frac{\Delta G^{xs,\alpha}}{RT} \qquad (6.13)$$

which is said to be applicable up to solubilities of 3 to 5 atomic percent. Calling the function on the left of Eq. (6.13) X_{corr}, the equation may be rewritten

$$X_{\text{corr}} = -\frac{\Delta H^{xs,\alpha}}{R}\frac{1}{T} + \frac{\Delta S^{xs,\alpha}}{R} \qquad (6.14)$$

It is readily seen that both Eqs. (6.3) and (6.13) may be used to obtain values of the enthalpies and excess entropies of solid solution from equilibrium solubility data. The method is illustrated for Eq. (6.14) in Fig. 6.7, where X_{corr} is plotted against the reciprocal of the absolute temperature for several alloy systems. The slopes of the resulting straight lines are $-\Delta H^{xs,\alpha}/R$, and the intercepts are $\Delta S^{xs,\alpha}/R$. The systematic deviations of the experimental data from the straight lines at low temperatures are characteristic of all such

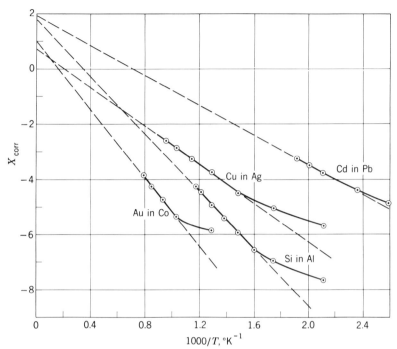

Fig. 6.7 *Solid solubility data plotted according to Eq. (6.14). (Data from Hansen[4.17] and the ASM "Metals Handbook.")*

data and emphasize the difficulty in obtaining true equilibrium at these temperatures. There is little doubt that the low-temperature solubilities given by the extrapolation of the straight lines in Fig. 6.7 are more nearly the true equilibrium solubilities than the experimental values.

6.5   PRIMARY SOLIDIFICATION IN EUTECTIC SYSTEMS UNDER EQUILIBRIUM CONDITIONS

The phenomena which occur during the cooling of alloys in a eutectic system may be most conveniently studied by reference to the five alloys of compositions 1 to 5 in Fig. 6.8. Alloy 1, representative of all alloy compositions between pure A and the point c in the figure, behaves exactly as if it were in a complete solid-solubility system such as that in Fig. 4.9. It begins solidification with the separation of α at the temperature where its alloy line crosses the

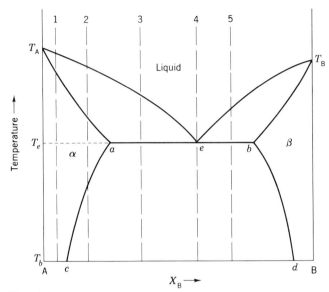

Fig. 6.8

liquidus line $T_A e$, finishes solidification at the intersection with the solidus line $T_A a$, and undergoes no further phase changes on continued cooling to the lowest temperature shown, T_b. This alloy "doesn't know" that it is in a eutectic system rather than one exhibiting complete solid solubility. These statements apply to alloys with compositions between pure B and the point d as well, with the one difference that the solid which forms is β rather than α.

The type of solidification represented by alloy 1 is called *primary* solidification. This is because certain alloys in binary systems, in this case those represented by alloys 3 and 5 in Fig.6.8, undergo two kinds of solidification: the first, or primary, solidification is identical with that which takes place in alloy 1; the second, or secondary, solidification takes place at the invariant temperature, as described later.

6.6   SOLID-SOLID PRECIPITATION UNDER EQUILIBRIUM CONDITIONS

Alloy 2, in Fig. 6.8, representative of all compositions lying between c and a, behaves just as if it were in a diagram containing a miscibility gap, such as those in Fig. 4.21. It starts solidification with the formation of α at the liquidus line $T_A e$ and finishes solidification at the solidus line $T_A a$; when, during further cooling, its composition line crosses the solvus ac, the B-rich solid solution β, of composition given at the intersection of the appropriate tie line with the solvus bd, begins to precipitate from the A-rich solid solution α. Continued extraction of heat is accompanied by further cooling and further precipitation of β, the remaining α becoming more A-rich and shifting its composition along the solvus ac. At the same time the newly formed β *and the previously formed β* shift in composition along the solvus bd by rejecting A atoms into the surrounding α and possibly also precipitating α within β particles. Which of the latter two ways of lowering its A content the β adopts is decided locally on the basis of two intrinsic rates. Consider, as in Figure 6.9, a β particle embedded in an α matrix. Consider further a region such as that at a within the β particle. This region can lower its A content in one of two ways: (1) it can reject the excess A atoms into the α matrix by diffusion over the distance l through the intervening β; or (2) the atoms in a part a' of this region can undergo the compositional and crystallographic rearrangements necessary to becoming an α particle, thereby lowering the A content of the remainder of the region a. The ease with which process (1) can take place is fixed by the rates of diffusion of A and B atoms through β, and by the distance l. The ease with which the second process can take place is determined by the rate at which an α particle of some particular finite size can form. Since both the diffusion rates and the nucleation rate of α particles are fixed largely by the temperature, at any given temperature the distance l determines whether process (1) or process

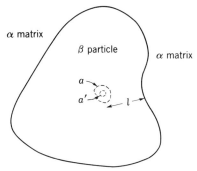

Fig. 6.9

(2) is more likely to occur. When l is small, that is, for small β particles or for regions near the surface of large β particles, process (1) tends to be favored, and for large l, that is, near the center of large β particles, process (2) tends to be favored. In practice the situation is not quite so simple as this, for the nucleation rate of one phase within another is highly sensitive to the presence of many kinds of tiny defect surfaces, e.g., those presented by second-phase impurities, grain boundaries, micropores, and microcracks. Formation of the new phase requires the production of an interface. Because of the transitional, and more or less chaotic, arrangement of atoms within this interface, it represents a region of relatively high free energy and is thus a barrier to the formation of the particle. Whenever part of this interface is presupplied, the formation of the new phase is easier (more rapid); this tends to happen at preexisting defect interfaces, where the atoms are "fooled" into "thinking" that the new phase with ready-made interface is already present. As a result the decision as to whether process (1) or process (2) above will in fact be chosen for alteration of the β-particle composition is modified by the presence of defect interfaces.

Returning now to Fig. 6.8, alloys with compositions lying between b and d undergo transformations on cooling qualitatively similar to those of alloy 2, with the roles of the α and β solid solutions reversed. It is clear, therefore, that all alloys in the eutectic system with compositions to the left of point a and to the right of point b behave as if they were part of the simpler alloy systems considered in Chap. 4. It is only for alloy compositions between a and b that manifestations of being in a eutectic system appear, for it is only these compositions which undergo the invariant reaction represented by the line ab.

6.7 EUTECTIC SOLIDIFICATION UNDER EQUILIBRIUM AND QUASI-EQUILIBRIUM CONDITIONS

Let us examine the nature of the eutectic invariant reaction by first considering the cooling of alloy 4 in Fig. 6.8. The unique feature of this alloy is that when it starts to solidify, the liquid is saturated with respect to both α and β simultaneously. The liquidus line $T_A e$ represents saturation of the liquid with respect to α, and consequently in alloys 1 and 2, and, as we shall see, in 3 as well, the solidification gets under way with the separation of α. Correspondingly, the liquidus line T_Be represents saturation with respect to β, and alloys on this side of the diagram first solidify by forming β. The alloy line for alloy 4, however, strikes both T_Ae and T_Be simultaneously at T_e; thus, alloy 4 may start its solidification with the formation of either α or β. In a large mass of liquid alloy 4 wherein a large number of solid nuclei may form at widely dispersed sites, under strictly equilibrium conditions some may be α and some β. These will have the compositions given by the points a and b, respectively, in Fig. 6.8. The reaction which takes place at T_e may be represented by the equation

$$L_e \rightleftharpoons \alpha_a + \beta_b \tag{6.15}$$

where the reaction to the right occurs on cooling and the reverse reaction on heating.

We have called this an invariant reaction, and it is easily predictable from the phase rule that this is so. This means, of course, that under equilibrium conditions the temperature and the phase compositions remain constant while the transformation is under way, i.e., as long as all three phases are present. It is instructive to consider from the viewpoint of the phase diagram why this is so. In Fig. 6.10 the eutectic diagram is redrawn to show the *metastable* extensions of the liquidus and solidus lines. These extensions have the following meanings. The extension *em* of the liquidus T_Ae represents compositions of liquid which would coexist with compositions of α along *ag* if a liquid in which α is forming were cooled below T_e under conditions of mutual equilibrium between α and liquid but in the absence of β; such conditions are called *metastable equilibrium*. (The metastable-equilibrium compositions may be found from free-energy curves by drawing the common tangents to the curves for the two phases present in metastable equilibrium; e.g., points corresponding to g and m in Fig. 6.10 may be obtained as the points of common tangency to the α and liquid curves at T_2 in Fig. 6.4.) Similarly, *eh* and *bn* represent metastable-equilibrium compositions between liquid and β. We now suppose that we have cooled the alloy of eutectic composition to $T_e - dT$ under equilibrium conditions and that at some point in the liquid an α particle is growing and at some other point a β particle is also growing. Equilibrium implies that the transformations are taking place so slowly that diffusion, relatively, is instantaneous. Thus the α, β, and liquid can be considered to be in instantaneous communication with each other

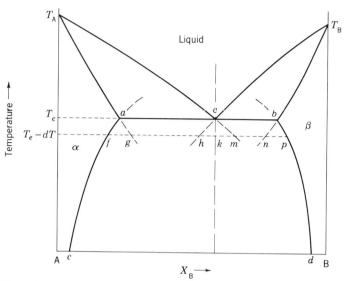

Fig. 6.10   *Eutectic diagram, showing metastable extensions of the liquidus, solidus, and solvus lines.*

even though the two solids may be separated by a finite distance. The liquid, to be in equilibrium with the α, will tend to adopt the composition m in Fig. 6.10, and the α the composition g. On the other hand, to remain in equilibrium with the β, the liquid must attain the composition h. These two values represent opposite shifts in the liquid composition from its original value e; the instantaneous interplay between the two tendencies will keep the composition at e and, thus, under equilibrium conditions, the temperature at T_e. Exactly analogous processes will keep the α composition at a and that of the β at b.

As has been indicated above, solidification under equilibrium conditions strictly speaking implies infinitesimally small, i.e., zero, rates. In any real solidification the rate must obviously be finite; this means that the solidification actually gets under way at some temperature a finite interval below the equilibrium temperature— never right at the equilibrium temperature! In more formal thermodynamic terms this is equivalent to saying that the driving energy of the process, or difference in free energy between the final state and the initial state, is zero at the equilibrium temperature* and

* The equilibrium temperature is, in fact, defined by this statement.

becomes finite only at some finite degree of undercooling. This driving energy is also essentially proportional to the degree of under-cooling. It is important to take into account this undercooling to understand the nature of eutectic solidification in real situations. We shall designate the near-equilibrium cooling conditions for which we shall examine eutectic solidification now as *quasi-equilibrium* to distinguish them from both strict equilibrium conditions and the more markedly nonequilibrium conditions which we shall consider later.

With reference to Fig. 6.11, which shows a portion of Fig. 6.8 enlarged, let us now imagine that liquid alloy 4 in Fig. 6.8 has been undercooled a finite amount to the temperature T_1 below T_e (Fig. 6.11). It is then represented by the alloy point k in Fig. 6.11. If a particle of α forms in the liquid, it will, as before, have a composition near that given by point a in Fig. 6.8; i.e., it will be far richer in A than the liquid from which it forms. The B atoms which the α particle rejects in forming must enter the immediately adjacent liquid, and under quasi-equilibrium conditions, this ambient liquid will maintain some increase in B content. Further growth of the α

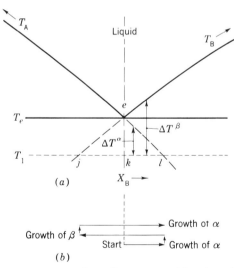

Fig. 6.11 *Phase-diagram considerations in eutectic solidification. (a) Undercooling of the liquid to k at T_1. (b) Shift in composition of the ambient liquid on the initial growth of an α region and the subsequent alternate growth of β and α regions.*

particle accentuates this excess B content in the ambient liquid. As a consequence, the composition of the liquid near the growing α particle moves, on the diagram in Fig. 6.11, from k toward l.* As it does so, the undercooling with respect to α, ΔT^α, decreases, since this undercooling is equal to the distance between the temperature horizontal jkl and the extension el of the liquid line $T_A e$ at the composition of the ambient liquid. At the same time the under-

* The temperature will also go up slightly due to the release of the latent heat of solidification. This effect does not alter the argument qualitatively.

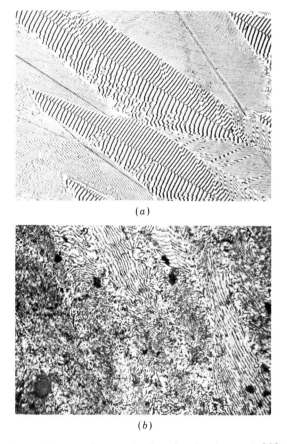

(a)

(b)

Fig. 6.12   *Some eutectic microstructures at 200×. (a) Ag–Sb; (b) Al–Cu, 18 atomic percent Cu; (c) Ag–Ge (white, massive regions are primary silver-rich solution); (d) Al–Si.   (Courtesy of L. F. Mondolfo.)*

cooling with respect to β, ΔT^β, which is equal to the distance between *jkl* and $T_B e$ at the ambient liquid composition, *increases*. This kind of undercooling, or increase in undercooling, brought about by a composition change rather than a temperature change, is called *constitutional undercooling*. As the α particle grows, then, because of the changes in constitutional undercooling accompanying the rejection of B into the surrounding liquid, the conditions become less conducive to the further growth of α and more conducive to the formation of β (at *l*, of course, the undercooling with respect to α would become zero, and the growth of α would no longer have any driving force). At some point in this shift of relative undercooling

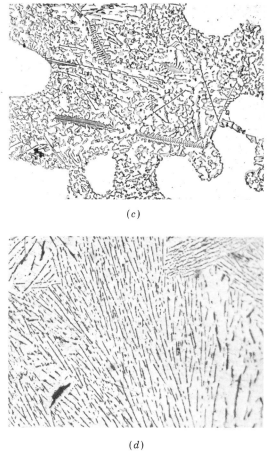

(*c*)

(*d*)

Fig. 6.12   (*Continued*)

the nucleation of a β particle in the ambient liquid becomes easier than the continued growth of α, and a β particle forms. Its formation, and growth, reverses the shift of constitutional undercooling, so that after a short period of growth an adjacent α particle forms and grows, etc. The result of this process is to produce a solidified eutectic product consisting of an intimate mixture of alternate regions of the two solid phases.

The shapes and interphase spacings of eutectic mixtures depend on many factors which are as yet only partly understood. Among these the most important are the degree of undercooling, the diffusion rates of the components, and the rate of heat dissipation—particularly in the liquid, the rates of nucleation at the interfaces of the growing phases, the interfacial energies, the crystal structures of the solid phases, the relative growth rates in different crystallographic directions, and the degree of purity. Some typical eutectic structures are shown in Fig. 6.12.* The property common to these structures is the systematically alternating regions of the two solid phases on a spacing scale usually small compared to the size of individual regions which form during primary solidification when the liquid is saturated with respect to one solid phase only (see especially Fig. 6.12c).

It is found experimentally that the two phases in eutectics characteristically form most often as alternating platelets, or lamellae, rather than as, say, alternating concentric spheres. The structures in Fig. 6.12a, b, and d are examples; the three-dimensional platelets appear in the two dimensional planes of microscopic observation as lines of greater or lesser thickness, the latter depending largely on the angle between the observation plane and the plane of the lamellae. The reason for the lamellar growth may be understood in terms of the sketch in Fig. 6.13. Let us suppose that an α particle (1) in Fig. 6.13 has nucleated first and that β particles (2) and (3) then form adjacent to the α. Along the interface between the liquid and the composite solid there will now be set up two-directional diffusion gradients. Near the α–liquid interface the liquid becomes B-rich and near the β–liquid interfaces A-rich. The resulting compositional gradients can be dissipated now not only into the liquid in a direction perpendicular to the solid–liquid interfaces but also by diffusion in the direction lateral to this, as indicated by the arrows in Fig. 6.13. This relieves the tendency toward nucleating

* For excellent treatments of microstructures and their interpretation, see Refs. 6.3 and 6.4.

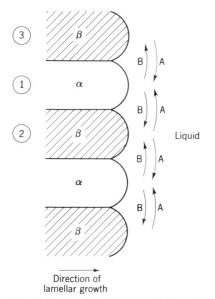

Direction of
lamellar growth

Fig. 6.13   *Lamellar growth in eutectic solidification.*

the opposite solid phase ahead of each growing solid region and allows the development of steady-state growth conditions perpendicular to the liquid–composite-solid interface. As a consequence the solid regions grow as lamellae with their long dimensions in this direction. The sidewise nucleation of alternating solid phases continues, of course, and the final product consists of *nodules* of lamellae (see particularly Fig. 6.12a), each nodule resulting from a separate initial nucleation event.

The characteristic nature of eutectic solidification products appears in practice even after quite slow cooling. It is, not, however, an equilibrium phenomenon since it requires for its formation a concentration buildup in the ambient liquid adjacent to a growing particle. This buildup can occur and be maintained only when the particle–liquid interface advance can supply excess A or B atoms fast enough to balance the rate at which diffusion can carry them away. The mobility of the atoms in diffusion is not, at a given temperature, dependent on the rate of heat extraction, but the velocity of the particle–liquid interface is; thus, at vanishingly small rates of heat extraction, i.e., equilibrium rates, the concentration buildup in the ambient liquid will disappear, and there will then be no reason for the alternating growth of the two solid phases in a particular locality. In this case the expected solidification mechanism would be very much like primary solidification for each particle which forms; there would remain the one difference that some of the particles would be α and some β, so that the resulting solid would consist of a relatively gross mixture of two essentially primary phases.

6.8   ALLOYS UNDERGOING BOTH PRIMARY AND EUTECTIC SOLIDIFICATION

Returning again to Fig. 6.8, alloys of composition 3 and 5 undergo both primary and eutectic solidification. In alloy 3 the primary phase is α, whereas that in alloy 5 is β. In both, however, the eutectic is the same and, as a matter of fact, is the same eutectic which forms in alloy 4. To demonstrate this, let us follow the cooling of, say, alloy 5 under quasi-equilibrium conditions. It starts solidification at the liquidus T_Be with the separation of primary β crystals. As cooling proceeds, the liquid shifts composition along T_Be, the β along T_Bb. When the temperature T_e is reached, the liquid has attained the composition e; it is now in every respect identical with the liquid in alloy 4—and for that matter in all alloys between a and b—at this temperature. This liquid, therefore, undergoes the same eutectic solidification at T_e as does that in alloy 4, and the product of the solidification is the same intimate mixture of alternating α_a and β_b regions. The microstructure will exhibit, in addition to this eutectic mixture, however, the more massive crystals of β formed during the primary solidification. Alloy 3 will contain the eutectic mixture and primary α crystals. Examples of typical microstructures in such alloys are given in Fig. 6.14. It should be particularly noted in these photomicrographs, and in those in Fig. 6.12, that the eutectic mixture frequently has one phase distributed within the other as a matrix. This usually happens when the eutectic composition is close to that of one solid phase; the latter phase, being present in the eutectic in greater amount (as may be shown by the lever rule), will tend to envelop the other solid phase. The matrix phase is then the same throughout the range of alloy compositions, as demonstrated in Fig. 6.14 for the Al–Si system. Here the eutectic composition is 11.6 percent Si, and accordingly the Al-rich solid solution is found to be the eutectic matrix; as a result, in alloys for

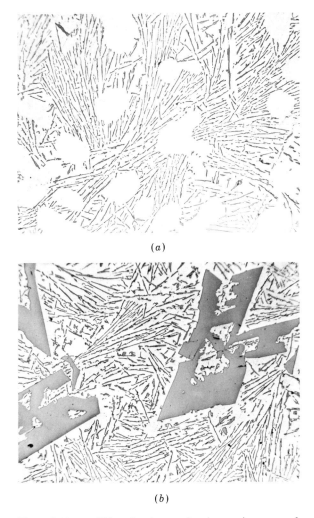

(a)

(b)

Fig. 6.14 *Microstructures showing primary and eutectic solidification in the Al–Si system. (a) Primary Al-rich crystals continuous with the Al-rich eutectic matrix. (b) Primary Si-rich crystals separated by inter-phase boundaries from the Al-rich eutectic matrix. 200×. (Courtesy of L. F. Mondolfo.)*

which primary Al-rich crystals appear along with eutectic (Fig. 6.14a) the primary phase and the eutectic matrix are continuous, whereas when the primary crystals are the Si-rich phase, there is a distinct interphase boundary between the primary crystals and the eutectic matrix (Fig. 6.14b).

6.9   SOME NONEQUILIBRIUM PHENOMENA IN EUTECTIC SYSTEMS

When alloys in a eutectic system are cooled at nonequilibrium rates, the solidification range is lowered and coring takes place for the same reasons as described in connnection with complete solid-solution systems (see Fig. 4.10). In addition, a number of other phenomena may take place; three frequently met examples are described below.

In an alloy in which both primary and eutectic solidifications take place but in which the primary phase is not the same as the eutectic matrix phase, the phenomenon called the *halo effect*, illustrated in Fig. 6.15a for the Ag–Cu system, may occur. Here, the liquid around the Cu-rich primary crystals (dark) was presumably able to support a large concentration gradient without nucleation of the Ag-rich phase (light), resulting in a broad region around each primary crystal in which the copper concentration was very low. When eutectic freezing reached these regions later, there was not enough copper in the liquid to produce Cu-rich particles in the eutectic, and thus each primary crystal tends to be surrounded by a halo free of Cu-rich eutectic particles.

A solidification structure called a *divorced eutectic*, which is perhaps an extension of the halo effect, may be produced under conditions of large undercooling, most readily in alloys of compositions such that the primary phase, say α in Fig. 6.16a, and the eutectic matrix phase are the same. If the β phase does not nucleate easily, the alloy may be brought without the formation of β down to the temperature at which the metastable α solidus and the alloy compositions coincide, point a in Fig. 6.16a. Under these conditions, the newly forming α will have the composition a, and the liquid around each α crystal will have the composition b in Fig. 6.16a, but the main mass of the liquid may still have its original composition, that of the alloy, a. It may be recognized that these conditions are exactly analogous to those which produce steady-state freezing during zone refining (see discussion in Sec. 4.9). As the solid-liquid interface advances (see sketch in Fig. 6.16b), the B atoms rejected from the α are delivered by diffusion into liquid of the same concentration as the α which is forming, and steady-state freezing of the α crystal sets in. Ultimately, when the concentration buildup in regions around individual α crystals impinge on each other, the B content in the liquid will rise just as it does at the end of a bar in zone

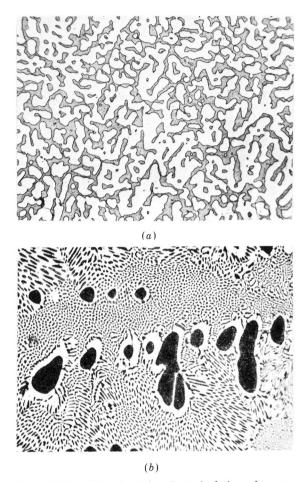

(a)

(b)

Fig. 6.15 (a) *The halo effect, depletion of copper from the eutectic regions adjacent to the Cu-rich primary crystals in an Ag–Cu alloy.* (b) *Divorced eutectic in a magnesium-rich Al–Mg alloy.* 225 ×. (*Courtesy of L. F. Mondolfo.*)

refining, and the β will finally nucleate in the high-B liquid. The resulting structure, however, does not have the typical eutectic appearance; an example of a divorced eutectic is given in Fig. 6.15b.

In certain dilute alloys, *false eutectic freezing* may occur. For example, the alloy of composition 1 in Fig. 6.17 does not undergo eutectic solidification when cooled at essentially equilibrium rates. As a result of coring in nonequilibrium cooling, however, the average composition of the solid may be at the point a in Fig. 6.17 when the

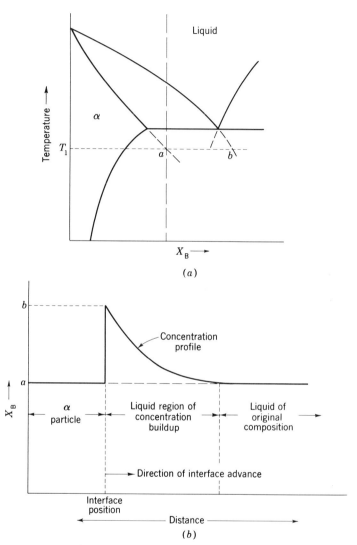

Fig. 6.16   *Conditions for divorced eutectic.* (a) *Alloy with primary α undercooled to* T_1 *without nucleation of β;* (b) *steady-state growth of the α crystals at* T_1.

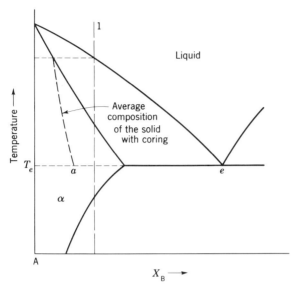

Fig. 6.17   *Conditions for false eutectic freezing.*

alloy has reached the eutectic temperature T_e. There will then be a
considerable quantity of liquid left, and the liquid will now be
saturated with respect to β as well as α; hence, eutectic freezing at
essentially constant temperature can now take place, giving the
phenomenon of false eutectic freezing.

6.10   THE EUTECTIC-LIKE SYSTEMS

As indicated before, three eutectic-like invariant reactions are known
to occur in binary alloy systems. Diagrammatically, the only differences between these and the eutectic invariant lie in the number
and placement of the liquid phases involved. This is illustrated in
Fig. 6.18. If one of the solid phases in the eutectic is replaced by a
liquid phase, the invariant known as the *monotectic* results (Fig.
6.18b). The invariant reaction may in this case be written

$$L_2 \rightleftharpoons \alpha_1 + L_3$$

where, as before, cooling is represented by the reaction to the right,
and heating to the left. If, on the other hand, the liquid phase in the
eutectic is replaced by a solid phase, so that now all three phases

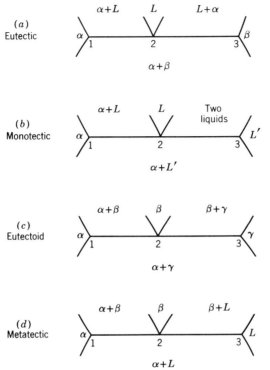

Fig. 6.18   *Symbolic representation of the eutectic and the eutectic-like invariant reactions.*

are solid (Fig. 6.18c), there results the eutectoid invariant. Here the equation for the reaction is

$$\beta_2 \rightleftharpoons \alpha_1 + \gamma_3$$

Finally, if the liquid phase and one solid phase in the eutectic are simply interchanged in position, the invariant shown in Fig. 6.18d results, with the equation

$$\beta_2 \rightleftharpoons \alpha_1 + L_3$$

This invariant reaction has as yet no generally accepted name but the designation *metatectic* is sometimes used.

The eutectic-like invariants do not usually appear as simple systems complete in themselves, as is frequently the case with the eutectic. Although it is possible to draw hypothetical diagrams of

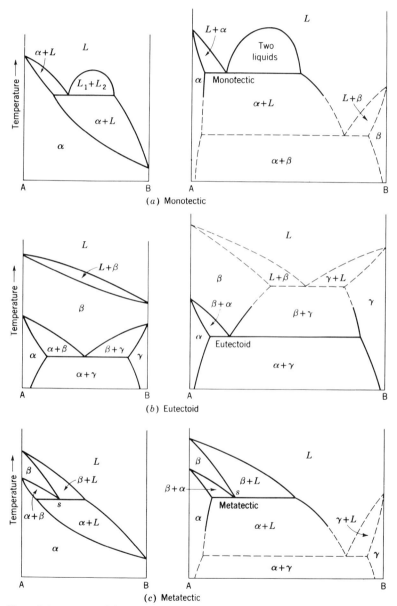

Fig. 6.19   *Possible diagrams incorporating (a) the monotectic, (b) the eutectoid, and (c) the metatectic invariant reactions.*

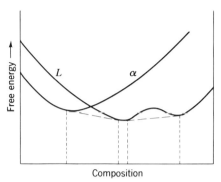

Composition

Fig. 6.20 *Schematic free-energy curves for liquid and solid just above the monotectic temperature in a simple monotectic system.*

this type, as illustrated in Fig. 6.19 on the left, in practice the eutectic-like reactions generally appear as portions of systems involving such complications as polymorphic transformations and/or other invariant reactions and intermediate phases. Schematic examples are given on the right in Fig. 6.19; real examples are the Cu–Pb, Fe–O, and SiO_2–MnO systems for the monotectic, the Fe–Ta, Fe–C, and MgO–ZrO_2 systems for the eutectoid, and the U–Mn system for the metatectic. The simple monotectic on the left in Fig. 6.19a does not occur because it is extremely unlikely that two components which are partially immiscible in the liquid state will exhibit complete solubility in the solid state; usually the solubility will be more limited in the solid state, as on the right in the figure. In thermodynamic terms, the simple monotectic would imply free-energy–composition curves which, for example, just above the monotectic temperature would be of the type shown in Fig. 6.20. The inflection in the liquid curve and lack of one in the solid would in turn imply either (1) a higher positive ΔH^{xs} in the liquid than in the solid or (2) a higher entropy in the solid than in the liquid. Both of these conditions are unlikely because liquids generally have more voluminous, "looser," and more random atomic structures than solids. There seems to be no theoretical reason, on the other hand, why the simple eutectoid or metatectic diagrams on the left in Fig. 6.19 should not be found. Their nonoccurrence is, perhaps, attributable simply to the statistical improbability of finding two components with the very restrictive set of characteristics involved—the required allotropic forms in the appropriate temperature ranges and the correct crystal structures coupled with the close chemical and physical atomic similarity necessary for complete solid solubility.

Of the three eutectic-like invariants the metatectic is of relatively rare occurrence and will not be discussed in detail; it is, however, somewhat of a curiosity in that certain alloys in the system can melt *on cooling!* For example, in Fig. 6.19c, any alloy of composition to the left of the metatectic composition *s* will be completely solidified when cooled to the metatectic temperature but will partially melt again on further cooling into the α-plus-liquid region just below this. By far the most frequently occurring of the eutectic-like invariants is the eutectoid. It is the basis of many important solid-state transformations; in particular, in the iron-carbon system a eutectoid transformation is responsible for the ability to produce steels with very high strengths and many other very useful properties.

In addition to the three eutectic-like reactions discussed above, the term *monotectoid* is sometimes met in phase-diagram work. This term is useful in designating a reaction having the form of the monotectic shown on the right in Fig. 6.19a but in which all three phases involved are solids. It is not listed here as a separate eutectic-like reaction, however, because the monotectoid and eutectoid are exactly alike in every way other than the appearance in the former of the closed-loop formation above the invariant line.

The types of free-energy curves which lead to eutectic-like invariants are illustrated in Fig. 6.21 for the eutectic-eutectoid combination shown in Fig. 6.19b. It is left to the student to work out similar curves for the diagrams in Fig. 6.19a and c.

The transformations which take place during heating and cooling alloys in eutectic-like systems are analogous *in principle* to those which take place in the eutectic system, so that the previous discussions pertaining to the latter are largely applicable to the former. There are certain practically important differences, however, which lead to microstructural differences; these arise largely from the fact that mass flow, both convective and diffusive, takes place orders of magnitude faster in liquids than in solids. For example, the product of eutectoid transformation on cooling gives rise to an intimate mixture of two solid phases in much the same way and for the same reasons as in eutectic solidification; however, the two phases in a eutectoid structure tend to be more regularly arrayed than in a eutectic structure because in the latter the presence of the liquid

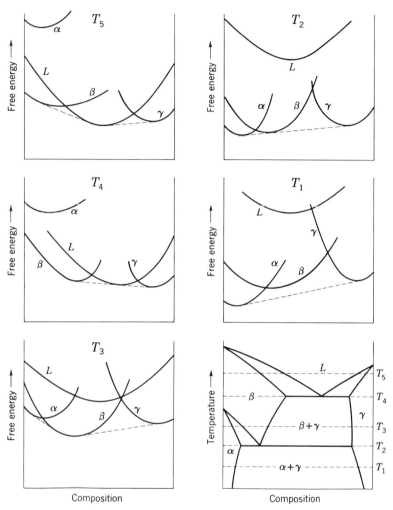

Fig. 6.21   *Schematic free-energy curves leading to phase diagram with a eutectic and a eutectoid invariant reaction.*

allows the growing solids to undergo accidental alterations in growth direction more freely. Further, the eutectoid mixture tends to be characterized by a smaller interlamellar spacing because the lateral diffusion at the advancing edge of the composite growing nodule (see Fig. 6.13) is limited to shorter distances by the relatively low diffusion rates in solids as compared with liquids. Typical eutectoid

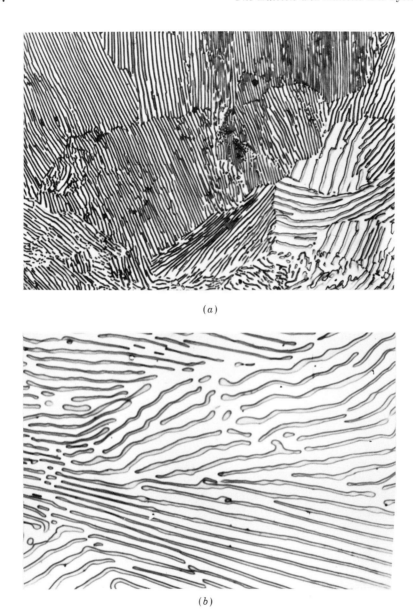

(a)

(b)

Fig. 6.22   *The microstructure of pearlite, (a) 1000×;*
(b) 2500×. (From Bain and Paxton, "Alloying Elements
in Steel," ASM. Courtesy J. R. Villela and H. C.
Knechtel, Edgar C. Bain Laboratory for Fundamental
Research, United States Steel Corp., Monroeville, Pa.)

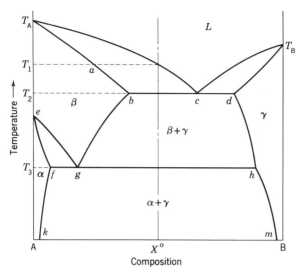

Fig. 6.23   *An alloy $X°$ which undergoes both a eutectic and a eutectoid transformation during cooling or heating.*

microstructures are shown in Fig. 6.22; these are representations of the eutectoid decomposition product in steel called *pearlite* because of the mother-of-pearl appearance it sometimes gives when viewed under a low-magnification microscope.

The monotectic invariant is found to occur most frequently in systems where the two components are significantly dissimilar electrochemically, as, for example, in metal-nonmetal systems, some nonmetal-nonmetal systems, and even in certain metal-metal systems where the two metals are far apart in the periodic table. This is to be expected because the occurrence of partial immiscibility in the liquid state, the *sine qua non* of the monotectic, should require atomic or molecular species of quite different physical and chemical properties. As a corollary, in monotectic systems the mutual intersolubility in the solid state tends to be very limited.

The monotectic transformation, just like the eutectoid, is analogous to the eutectic; in this case it differs only in that one of the end-phases is liquid rather than solid. In all known monotectic systems, however, the monotectic composition occurs very close to that of the solid end-phase. As a result, as can be seen by application of the lever rule, the monotectic decomposition product during cooling consists largely of solid; accordingly, the liquid which forms during the monotectic transformation appears as small pockets surrounded by the solid phase. This liquid ultimately solidifies during cooling, of course—in the case shown in Fig. 6.19a by a eutectic transformation; in an alloy of the monotectic composition this leads to a final microstructure consisting of islands of a fine eutectic mixture embedded in a matrix of the solid phase formed by the monotectic reaction (see Ref. 6.1 for examples).

As an exercise in the use of diagrams containing more than one invariant reaction, the quasi-equilibrium cooling of one alloy in a system such as that in Fig. 6.19b will be followed in detail. The diagram is re-presented in Fig. 6.23. All those alloys—and only those alloys—whose overall compositions lie between the points b and d in Fig. 6.23 will go through both invariant reactions. One such alloy is that having the composition $X°$ in the figure. This alloy begins its solidification during cooling at the temperature T_1 with the separation of β solid solution of composition a. As heat extraction proceeds, solid β continues to form, its composition shifting along the line ab and that of the B-enriched liquid shifting along the liquidus $T_A c$. At T_2 the constant-temperature eutectic reaction takes place, the remaining liquid of composition c transforming to more β of composition b and γ of composition d. When solidification is complete, further heat extraction causes the temperature to decrease again, because, as the phase rule predicts, with only two phases present once again a degree of freedom exists. Thus, just below T_2, the alloy consists of solid β in two forms—the relatively massive β produced by the primary solidification between temperatures T_1 and T_2 and the more finely dispersed β in the eutectic mixture—plus solid γ in the eutectic.

The proportion of the alloy now existing as primary β and as eutectic can readily be predicted from the lever rule. The fraction of eutectic is equal to that portion of the alloy which was liquid just before the eutectic reaction set in and thus is

$$\text{Eutectic} = \frac{X° - b}{c - b}$$

The fraction of primary β is, correspondingly,

$$\text{Primary } \beta = \frac{c - X°}{c - b}$$

As cooling continues through the $\beta + \gamma$ region, the γ shifts its composition along the line *dh*, and the β, both the primary and the eutectic β, along the line *bg*. These composition adjustments take place by simple diffusional processes and/or by solid-state precipitation, as described previously. Of course, the relative amounts of β and γ change also, so that at a temperature just above T_3 the proportions correspond to those given by the lever rule as applied to the lever extending from *g* to *h* with its fulcrum at $X°$. At T_3, the β composition now may be seen to lie not only on the line *bg* but also on the line *eg*; the β is, therefore, now simultaneously saturated with respect to both α and γ, and further heat extraction causes it to undergo eutectoid decomposition, forming α of composition *f* and more γ of composition *h*. This happens to both the β which formed during primary solidification and the β existing as part of the eutectic. The γ in the eutectic does not take part in this decomposition; it simply "awaits" the completion of the eutectoid transformation. When this has happened, the alloy consists only of α of

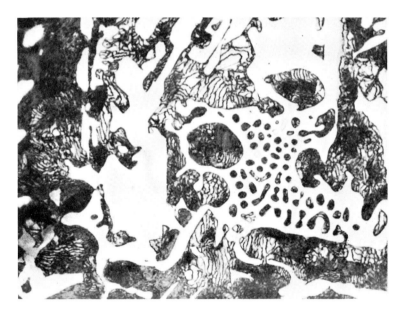

Fig. 6.24   *The microstructure of "white" cast iron containing approximately the eutectic carbon content. Large white regions are eutectic cementite (iron carbide); remainder is pearlite resulting from the eutectoid decomposition of the austenite formed during solidification in the eutectic region. 1000×.*

composition f and γ of composition h, and during subsequent cooling these phases undergo final compositional adjustments along the lines fk and hm. Though the alloy, thus, ultimately contains only the phases α and γ, the various transformations will have their effects on the final microstructure through the various forms the two phases will have adopted, i.e., on the microconstituents which will be present. Relatively large areas of eutectoid structure will exist in

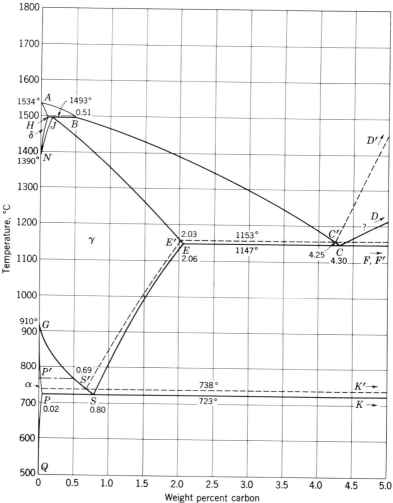

Fig. 6.25 *The iron-carbon phase diagram. (From Hansen.[4.17])*

the regions which were originally primary β. The remainder of the microstructure will consist of the eutectic mixture of γ regions interspersed with eutectoid regions which were formerly eutectic β, the eutectic particle size being intermediate between that of the primary β and that of the finely divided eutectoid lamellae. A "white" cast iron has a microstructure of this type, as illustrated in Fig. 6.24. This microstructure results from the fact that the Fe–C diagram in the Fe-rich region (Fig. 6.25) corresponds very closely to the schematic diagram of Fig. 6.23.*

Problems

6.1   It is found in the zone refining of aluminum that the impurity copper is removed much more efficiently than the impurity silicon. To indicate the reason for this, make calculations using Eq. (6.2). Compare your calculated quantities with the similar quantities found from phase-diagram data alone in Prob. 4.8.

6.2   Show that for different solutes in a given solvent the slope of the liquidus line near the melting point of the pure solvent (a) is independent of the nature of the solute if the solubility of the solute in the solid solvent is small compared to that in the liquid and (b) varies linearly with the ratio of solid to liquid solubility of the solute if the two solubilities are not too different. (Assume that the liquidus and solidus are essentially straight lines in this region near the melting point of the pure solvent.)

6.3   Show that for alloy systems of solvents which follow Richard's Rule (see Prob. 3.13) the slope of the liquidus line near the melting point of the pure solvent is not only independent of the nature of the solute (as in Prob. 6.2) but is proportional to the absolute melting temperature of the solvent if the solid solubility of the solute is small compared to the liquid solubility (again assume the liquidus and solidus are straight lines near the melting point of the solvent).

6.4   Test the validity of Probs. 6.2a and 6.3 with published phase-diagram data for several metal solvents.

---

* The upper left-hand corner of the Fe–C diagram exhibits a peritectic invariant. (see Chap. 7). The alloy for which the microstructure in Fig. 6.24 is given contains enough carbon so that it does not encounter the peritectic invariant. It should be noted, however, that this peritectic invariant occurs at such high temperatures that even alloys of carbon content low enough to go through the peritectic invariant do not normally exhibit evidence of this in their ultimate microstructures. This is because at these high temperatures subsequent solid-state diffusion is rapid enough to effectively eliminate all traces of the peritectic transformation effects.

6.5   The solid solubility of copper in aluminum is 1.7 atomic percent at 500°C and 0.68 atomic percent at 400°C. What is it at 200°C?

6.6   For equilibrium cooling of alloy 3 in Fig. 6.8, calculate in percent of the total weight of the alloy (a) the amount of eutectic at $T_e - dT$; (b) the amount of α in the eutectic at $T_e - dT$; and, (c) the amount of β precipitated from the primary α during cooling from T_e to T_b, assuming that each α particle adjusts its composition only by the precipitation of β within it.

6.7   Describe as completely as you can the changes in alloy 5 (Fig. 6.8) during slow cooling and heating.

6.8   Repeat Prob. 6.7 for an alloy composition which passes through the two-liquids region in the system shown at the right in Fig. 6.19a.

6.9   Repeat Prob. 6.7 for an alloy composition lying between the eutectic composition and the maximum γ solid solubility in the system shown at the right in Fig. 6.19b.

6.10   Repeat Prob. 6.7 for an alloy composition lying between the metatectic composition and the maximum α solid solubility in the system shown at the right in Fig. 6.19c.

6.11   Draw schematic free-energy–composition curves at several representative temperatures for the alloy system at the right in Fig. 6.19c.

6.12   As stated earlier, graphite rather than Fe_3C is the stable carbon-rich phase in the Fe–C system. Note that in published Fe–C diagrams, solubility lines for both are often given. For example, the compositions of γ-Fe in equilibrium with both Fe_3C and graphite are given, that for Fe_3C lying to the carbon-rich side of that for graphite. Show by means of schematic free-energy–composition curves for the system that this is the proper relation for the two solubilities.

References

6.1   F. N. Rhines, "Phase Diagrams in Metallurgy," McGraw-Hill Book Company, New York, 1956.

6.2   J. F. Freedman and A. S. Nowick, *Acta Met.*, **6**:176 (1958).

6.3   W. Rostoker and J. R. Dvorak, "Interpretation of Metallographic Structures," Academic Press Inc., New York, 1965.

6.4   R. M. Brick, R. B. Gordon, and A. Phillips, "Structure and Properties of Alloys," 3d ed., McGraw-Hill Book Company, New York, 1965.

SEVEN

TWO-COMPONENT SYSTEMS CONTAINING INVARIANT REACTIONS: THE PERITECTIC AND PERITECTIC-LIKE SYSTEMS

7.1 THE PERITECTIC SYSTEM

A simple schematic system consisting of two solid phases and a liquid phase which take part in a peritectic invariant reaction is shown in Fig. 7.1. The diagram looks significantly different from a eutectic, and in point of fact leads to microstructures which are appreciably different from those in a eutectic system, as we shall see. From the viewpoint of the free-energy curves which lead to the invariant reactions, however, the difference between the two types of diagrams is not large. Typical free-energy curves for the peritectic are shown in Fig. 7.1. Comparing these curves with those for the eutectic in Fig. 6.4, it is seen that the major difference lies in the relative positions assumed by the minima in the free-energy curves along the composition axis as the temperature of the invariant is approached. In the eutectic, the liquid minimum lies between those of the two solids, whereas in the peritectic it lies outside those of the solids. The latter situation is favored by a large difference in the melting points of the components.

It was noted before that there is a close similarity between the eutectic phase diagram and one characterized by a liquidus-solidus minimum, continuous solid solution at high temperatures,

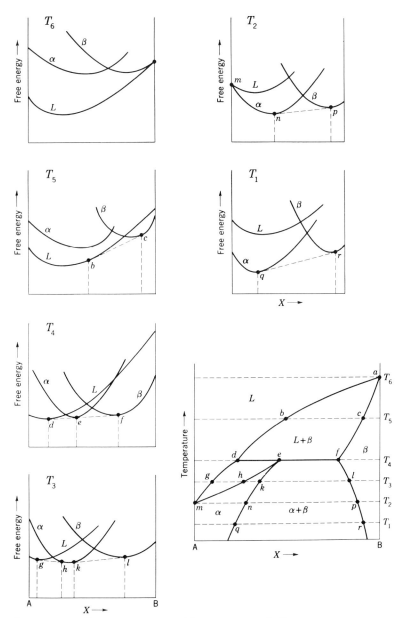

Fig. 7.1   *Free-energy–composition curves and the temperature-composition equilibrium diagram for a peritectic system.*

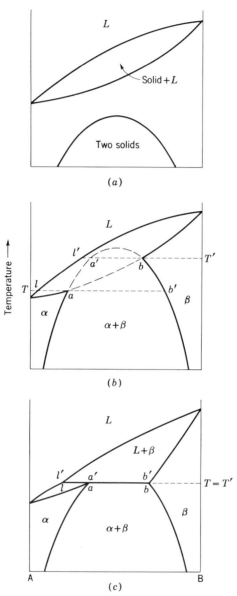

Fig. 7.2   *Similarity between a peritectic diagram and a diagram exhibiting a solid miscibility gap but no minimum or maximum in the liquid-solid region.*

and a solid solubility gap at low temperatures (see Figs. 4.21c and and 6.5). In an analogous fashion the peritectic diagram is closely related to one showing a solid solubility gap at low temperatures but having no minimum in the liquid-solid region (Fig. 4.21b). A diagram of the latter type is shown in Fig. 7.2a. In Fig. 7.2b a qualitatively identical diagram is depicted, but it has been supposed that $G^{xs,S}$ is now enough larger than $G^{xs,L}$ so that the miscibility gap extends upward to temperatures where liquid is still stable; i.e., the miscibility gap and the liquid-solid region intersect. Just as for the eutectic (see discussion in Sec. 6.2), this leads to an invariant horizontal in the diagram (Fig. 7.2c). Since liquid, α, and β of compositions l, a, and b' are in equilibrium at T and of compositions l', a', and b are in equilibrium at T', the points l-l', a-a', and b-b' must coincide, respectively; this follows from the phase rule, which states that equilibrium between a given three phases in a binary system at arbitrarily fixed pressure can occur at only one temperature and involve only one specific composition of each phase.

7.2   EQUILIBRIUM SOLIDIFICATION IN THE PERITECTIC SYSTEM

Referring to the typical peritectic system in Fig. 7.3, it may readily be seen that alloys 1, 2, 6, and 7 undergo transformations during cooling which are exactly analogous to those in alloys 1 and 2 in the eutectic system of Fig. 6.8. These will, therefore, not be discussed in detail; it should be noted, however, that the primary solidification in alloys 1 and 2 (Fig. 7.3) produces α, as does that in all alloys with B content less than point c in the diagram, whereas primary solidification in alloys 6 and 7, and all those with compositions greater than point c, produces β. None of these four alloys undergoes the peritectic invariant transformation, but in alloys 3, 4, and 5 (and all others with compositions between points a and c) the peritectic transformation is encountered. In each of these alloys the primary phase is α, and primary solidification proceeds with a lowering of temperature accompanied by a shift in the α composition along the solidus line $T_A a$ and the liquid composition along the liquidus line $T_A c$. When the peritectic temperature T_p is reached, the liquid composition has attained the value c; the liquid is now saturated with respect to the second solid phase β as well as with α, since its composition lies both on cT_B and $T_A c$. Thus, solid

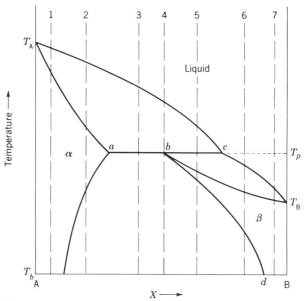

Fig. 7.3

β now forms, and with the appearance of the third phase the reaction becomes invariant. The transformation which takes place here is quite different from that in a eutectic system; it consists of a reaction between the primary solid and the liquid to produce the second solid. The reasons for this will be discussed in the next section. The peritectic reaction may be represented by the equation

$$L + \alpha \rightleftharpoons \beta$$

The only differences between the alloys 3, 4, and 5 in Fig. 7.3 are stoichiometric in origin. Application of the lever rule reveals that just above the peritectic temperature T_p the ratio of liquid to solid is low in alloy 3, higher in 4, and highest in 5. The alloy 4, in fact, has exactly the right proportions of α and L so that both are completely consumed in producing β, and after the peritectic reaction is over only β remains. This means that on subsequent cooling the alloy lies in the one-phase β region, but only momentarily in the particular case shown, the slope of the solvus line *bd* being such as to cause precipitation of α from the β on continued cooling. (It is worth noting that there is no theoretical reason why the line *bd* could not have the opposite slope, in which case cooling of the alloy 4 below T_p would keep it in the β region.) In alloy 3 there is an excess of β at a temperature just above T_p, and thus when the liquid has been consumed in producing β during the peritectic reaction, some α still remains. This alloy, consequently, enters the $\alpha + \beta$ region on further cooling. The reverse is true for alloy 5; it contains an excess of liquid, and completion of the peritectic reaction finds it in the $\beta + L$ region. Solidification is then completed in alloy 5 during further cooling through this region by a primary separation of β.

7.3   THE PERITECTIC REACTION

Because of the nature of the peritectic reaction, the product of the reaction has an entirely different microstructural appearance from that of the eutectic. For example, the microstructure of alloy 3 in Fig. 7.3 would consist of islands of primary α surrounded by a rather broad network of β. A typical example is the alloy in Fig. 7.4. There is no tendency for a finely divided, alternate-layer mixture of the two new phases to form, as is the case with the eutectic. Let us see why this is so. Consider the diagram in Fig. 7.5, which is a portion of the peritectic diagram in Fig. 7.3, and imagine that, say, alloy 4 has been cooled to the peritectic temperature T_p under virtual equilibrium conditions. If the liquid of composition c is undercooled to point c' at the temperature $T_p - dT$, it will now be undersaturated with respect to both α and β, since c' lies below both the liquidus *ecf* and the liquidus *ch*. We can now show that a layer of β will tend to form and grow continuously on the primary α present in this liquid. Let us suppose first that a primary α particle simply continues to grow at the expense of the liquid. As it does so, it rejects B into the ambient liquid, moving the liquid composition from c' toward f in Fig. 7.5 and its own composition from r' toward n. As this happens, the undersaturation of the liquid with respect to α decreases toward zero, tending to stop the growth of α, and also the α increases its oversaturation with respect to β. Both of these trends promote the formation of a layer of β between the α and the liquid, first because the liquid is still undersaturated with respect to β and second because formation of β gives the α a means of rejecting B atoms to reduce its B content toward p, the saturation value with respect to β. When the β layer forms, it will tend

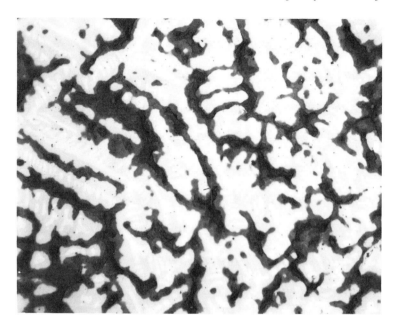

Fig. 7.4   *Typical microstructure of a cast peritectic alloy; Pt + 60% Ag. 850×. The white and light gray areas are cored primary solid solution; the dark is the second solid which formed during and after the peritectic reaction. (From Rhines.*[6.1])

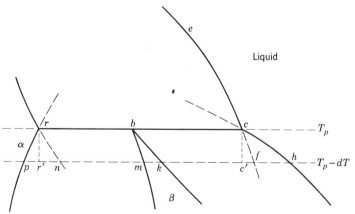

Fig. 7.5   *Peritectic diagram, showing metastable extensions of the liquidus, solidus, and solvus lines.*

to have the composition m on its α interface and k on its liquid interface. This produces a diffusion current of B atoms through the β from the liquid toward the α, which in turn *tends* to raise the B content in the β above m on the α side and to lower it below k on the liquid side. Readjustment of these concentrations then takes place by the β growing at the expense of both the α and the liquid; at the β-α interface, formation of more β from the low-B α requires the supplying of B from the already existing β, thus lowering its B content again toward m; at the β–liquid interface the attempt to raise the B content of the β toward k reduces the excess of B in the ambient liquid and thus tends to promote the further growth of β into the liquid. There is no further tendency for α to form, however, since there is no point in the process where the above tendencies reverse themselves, as they do in the case of the eutectic (see Sec. 6.7). The microstructure developed by the peritectic reaction thus becomes one in which just below T_p the two phases present are mixed on a size scale set by the size of the primary particles rather than by a new and finer subdivision produced, as in the eutectic, by the invariant reaction.

7.4   NONEQUILIBRIUM SOLIDIFICATION IN THE PERITECTIC SYSTEM

The peritectic transformation is a particularly sluggish phenomenon. As pointed out several times, the maintenance of equilibrium requires sufficient time for composition adjustments by diffusion. In eutectic solidification these adjustments take place largely by diffusion through the liquid phase (see Fig. 6.13); in the peritectic transformation, however, diffusion must take place through the solid phase being formed, for, as described above, this phase forms as a layer separating the two reacting phases. Since diffusion in solids is inherently slow, the peritectic reaction is also slow and is thus easily suppressed by even moderate cooling rates. As a matter of fact, special efforts frequently have to be made to allow the reaction sufficient time to go to completion. If this is not done, the nonequilibrium phenomenon called *envelopment* will take place. With reference to the diagram in Fig. 7.6a, the alloy of composition 1 should, after equilibrium cooling, consist only of β solid solution. If, however, the alloy is cooled too rapidly, the first formation of β around the primary α will effectively isolate the α from further

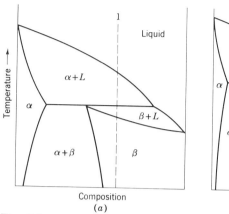
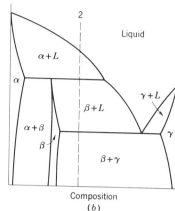

Fig. 7.6

contact with the liquid, and the alloy will then behave as if only the liquid and β were present, i.e., the α will be "enveloped" by the β. Under these circumstances, the remaining liquid will solidify by primary transformation to β, as in the $L + \beta$ region, and the final solidified product will consist not only of β but also of the trapped primary α regions which were prevented from reacting with the liquid.

In somewhat more complex systems, such as that in Fig. 7.6b, envelopment can lead to the presence of three phases in the solidified two-component alloy, whereas not more than two are normally expected. For example, in the alloy of composition 2 in Fig. 7.6b, equilibrium cooling would result in an alloy consisting of β and γ. If cooling is too rapid, however, primary α regions may become trapped by peritectic β, and the final structure will then include these islands of α as well as the expected β and γ.

7.5   THE PERITECTIC-LIKE SYSTEMS

Two peritectic-like invariant reactions are known to occur in alloys; these are the *peritectoid* and the *syntectic*. Although the peritectic is a very common reaction, both the peritectoid and syntectic are relatively rare. The peritectoid invariant is exactly analogous to the peritectic with the exception that all the phases involved are solid. If, for example, in the diagram in Fig. 7.3 the liquid solution were instead a solid solution, a peritectoid would result, with the reaction equation

$$\alpha + \gamma \rightleftharpoons \beta$$

The peritectoid may also involve intermediate phases and in practice does so more often than not. A schematic representation of such a diagram is given in Fig. 7.7; real examples may be seen in the Co–W, Ni–Mo, and FeO–Al$_2$O$_3$ systems.

Peritectoid reactions are even more sluggish than peritectic reactions, for they not only involve no liquids, but also they tend to occur at relatively low temperatures, where diffusion rates in the solids involved are extremely slow. For this reason it is usually impossible to produce complete transformation in peritectoid reactions, and frequently they are completely suppressed in the normal heating and cooling of alloys in which they occur.*

The syntectic, shown schematically in Fig. 7.8, is even of rarer occurrence than the peritectoid, and little is known of the resulting microstructures. The best known example occurs in the Na–Zn system. In general, for normal cooling rates, when the two liquids

* The recent work of E. R. Friese and others has indicated, however, that peritectoid reactions may sometimes proceed rapidly by martensite or martensite-like transformations.

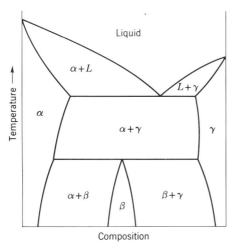

Fig. 7.7   *Diagram containing a phase (β) which forms by a peritectoid reaction.*

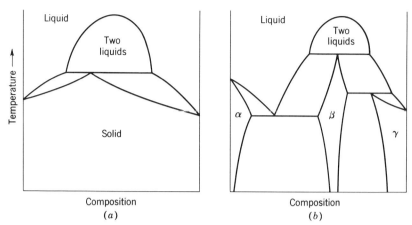

Fig. 7.8   *Two diagrams containing the syntectic invariant reaction.*

in the miscibility gap react to form the solid at the syntectic temperature, the layer of solid may be expected to isolate the liquids from each other in subsequent cooling. As a result the final microstructure will tend to be a coarse mixture of the structures characteristic of the lower-temperature reactions in the system. For example, in Fig. 7.8a the final microstructure of an alloy which undergoes the syntectic reaction will simply be a cored mixture of different compositions of the same solid solutions. In Fig. 7.8b, a syntectic alloy will ultimately show a mixture of the eutectic microstructure produced by solidification of the liquid of lower B content and the peritectic microstructure resulting from solidification of the B-rich liquid.

Occasionally the term *syntectoid* may be conveniently used to describe a peritectic-like reaction which has the form of the reaction in Fig. 7.8a but involves only solids. This reaction is not given separate status here, however, because it is identical with a peritectoid reaction in every respect other than the appearance of the closed loop above the invariant line.

Problems

7.1   Show in Fig. 7.3 that at T_p, under equilibrium conditions, the relative quantities of α and liquid in alloy 4 are exactly right to produce β with no excess of either α or liquid.

7.2 Describe the transformations in alloy 2 in Fig. 7.6*b* under (*a*) equilibrium cooling and (*b*) more rapid cooling.

7.3 For the system in Fig. 7.7 describe the microstructures one would expect at the lowest temperature shown if an alloy of a composition lying about halfway between the compositions of α and β at the peritectoid temperature were cooled from the liquid state at (*a*) a quasi-equilibrium rate and (*b*) an appreciably faster rate.

7.4 Draw schematic free-energy–composition curves at several representative temperatures for the system in Fig. 7.7.

7.5 Repeat Prob. 7.4 for the system in Fig. 7.8*b*.

7.6 Repeat Prob. 7.3 for the system in Fig. 7.8*b* and an alloy of composition equal to the composition of β at the syntectic temperature.

EIGHT

COMPLEX SYSTEMS

8.1 RULES FOR THE CONSTRUCTION OF COMPLEX PHASE DIAGRAMS

Real phase diagrams, even binary diagrams at fixed pressure, are very often quite complex, as a glance through such compendia of phase diagrams as M. Hansen and R. Elliott's, "Constitution of Binary Alloys," the American Society for Metals' "Metals Handbook," or the American Ceramic Society's "Phase Diagrams for Ceramists" will quickly show. A study of any of these complex diagrams also reveals, however, that each is just a combination of two or more of the simpler phase diagrams used as building blocks. Because phase diagrams must obey the laws of thermodynamics, certain principles may be applied in putting these building blocks together correctly. These principles may be summarized in a number of guiding rules, listed and illustrated in the discussion below.

1 The Phase Rule. *No construction in a diagram may violate the phase rule, $f = c - p + 2$.* Since we have already discussed the phase rule in some detail, it will not be further discussed here, but its use will be illustrated presently by a few examples.

2 The Boundary Rule. To present this rule, a few general characteristics of phase diagrams must first be noted.* We shall define the

* These characteristics apply to *all* phase diagrams, not only to the one- and two-component diagrams discussed here.

following quantities:

p = the number of phases in a phase region (as in the phase rule)

m = the number of variables (P, T, and compositions) plotted in the diagram

n = any integer, $0 < n \leqq m$

In terms of these quantities,

a   All diagrams are m-dimensional.

b p may have all values from unity to $m + 1$. The phase regions with $p = m + 1$ are the invariant regions.

c   All phase regions, with the exception of the invariant regions, have m dimensions. The invariant regions have $m - 1$ dimensions; they may, however, be looked upon as limiting regions in the same sense as, say, the limiting configuration of a triangle with gradually diminishing altitude is a straight line. For example, in a two-component system at fixed pressure, the three-phase invariant region is a line; however, this line is in reality three tie lines, each of which connects the compositions of two of the three phases in equilibrium. The three composition points could be considered to lie at the corners of a triangle which has an altitude of zero length. In fact, if even a small quantity of a third component were present, the composition tie lines would actually form a triangle with, now, an altitude of finite, but small, length. When the quantity of the third component is reduced to zero, as in a two-component system, the triangle degenerates to a line. In this sense, then, the invariant regions in all diagrams can be considered degenerate regions of m, rather than $m - 1$, dimensions, and *all* regions in a diagram can be said to have m dimensions.

Now, defining a phase-region boundary as an intersection between phase regions which has $m - 1$ dimensions, i.e., one fewer than the diagram itself, and considering for this purpose that the invariant regions are degenerate regions of m dimensions, the *boundary rule* can be stated as follows:

Any p-phase region can be bounded only by regions containing
$p \pm 1$ *phases*

This means, for example, that in a two-dimensional diagram such as that of a two-component system at arbitrarily fixed pressure, boundaries are one-dimensional, i.e., lines, and that two-phase regions can be bounded only by one-phase or three-phase regions, one-phase regions only by two-phase regions, and three-phase regions (the invariant regions) only by two-phase regions.

The boundary rule can be made more general, if desired, by allowing boundaries to be defined not only as intersections having one fewer dimensions than the diagram, but also as those having two fewer, three fewer, etc. In this more general way the boundary rule can be stated as follows:

For boundaries of n dimensions, any p-phase region can be bounded only by regions containing $p \pm (m - n)$ phases

The most useful form of the boundary rule is, however, the less general one given first; it may be seen that the second form reduces to the first simply by defining a boundary as a region having one-fewer dimensions than the diagram $(m - n = 1)$.

3   The Boundary-curvature Rule.   In two-component systems a useful rule with respect to the curvatures of one-phase region boundaries may be stated as follows (boundaries here are taken in the more limited sense of intersections having $m - 1$ dimensions, i.e., lines):

Boundaries of one-phase regions must meet with curvatures such that the boundaries extrapolate into the adjacent two-phase regions

This may readily be seen to be true by reference to the phase diagram and free-energy curves in Fig. 8.1. At temperature T_1 in this system liquid is metastable with respect to the stable phases α and β. In any two-phase alloy the stable condition would be α of composition e in equilibrium with β of composition k as given by the common tangent ek to the α and β free-energy curves. If liquid is caused to exist in metastable equilibrium with α in the absence of β at this temperature, the α and liquid compositions would be given by the points of common tangency c and g on the α and liquid free-energy curves. Since this combination of phases has a higher free energy than the stable combination, the tangent cg must lie above the tangent ek, and thus the point c must lie to the right of point e; that is, *the metastable solubility of a component in a phase is higher than the stable solubility.* Since c lies on the extrapolation of the one-phase boundary ab and lies in the two-phase $\alpha + \beta$ region, ab must extrapolate into this two-phase region. It can similarly be shown that db extrapolates into the two-phase region $\alpha + L$, and likewise for all the other one-phase boundaries in the system. Another way of stating this same rule is that *the included angle within a one-phase region between two boundaries of the region must be less than 180° at the intersection of the boundaries.*

4   The Solubility Rule.   *All components are soluble to some degree in all phases.* This rule was introduced in Chap. 4. As pointed out there, the extent of solubility is frequently so small as to be less than the width of a line on a chosen scale of composition, and, thus, in practice, diagrams may show phases with zero apparent solubility. In drawing schematic diagrams, however, the student should indicate solubility for all phases.

5   Use of Rules.   Some examples of common errors of construction are given in Fig. 8.2. Referring to the numbers and letters in the figure,

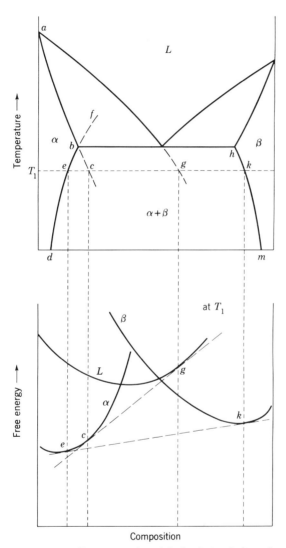

Fig. 8.1   *Demonstration of the basis for the boundary-curvature rule.*

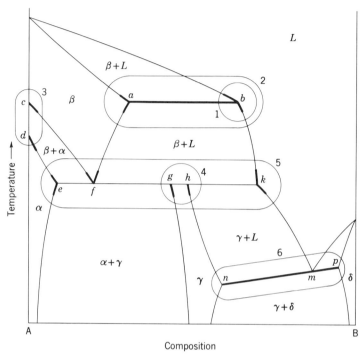

Fig. 8.2   *Some common errors in the construction of phase diagrams. Errors at (1) to (6) are discussed in text.*

the manner in which the errors violate the above rules is as follows:

Error 1   At the point *b* the boundary-curvature rule is violated.

Error 2   Along the line *ab* there are only two phases, β and *L*, indicated as being in equilibrium. The line *ab* is, thus, apparently a boundary, not an invariant region, and and should therefore not be horizontal since, by the phase rule, there is a degree of freedom along such a line. Further, *ab* violates the boundary rule since two two-phase regions cannot bound each other.

Error 3   The two-phase region β + α extends over the temperature range *cd* at pure A. This violates the phase rule, for, in a pure component, equilibrium between two phases *at arbitrarily fixed pressure* is invariant and, thus, can occur only at one temperature. If, alternately the β + α region had been marked as a single phase region to obviate this difficulty, the boundary rule would be violated,

for then one-phase regions would bound each other along
de and *cf* and a one-phase region would bound a three-
phase region (α-β-γ) along *ef*.

Error 4   The line *fghk* is presumably a three-phase equilibrium
and thus invariant. The range of compositions *gh* for the
γ phase in this equilibrium violates the phase rule, for in
an invariant reaction each phase can have only a single
composition. The boundary rule is also violated, for
along *gh* the one-phase γ region bounds the three-phase
$\beta + \gamma + L$ region.

Error 5   Four phases, α, β, γ, and L, are indicated as being in
equilibrium along the line *efghk*. This violates the phase
rule, which gives three as the maximum number of
phases which may coexist in equilibrium in a two-component system at arbitrarily fixed pressure.

Error 6   The line *nmp* violates the phase rule. Since it represents
three-phase equilibrium, and is thus invariant, it must
be a constant-temperature line.

A corrected form of the diagram in Fig. 8.2 is shown in Fig. 8.3.

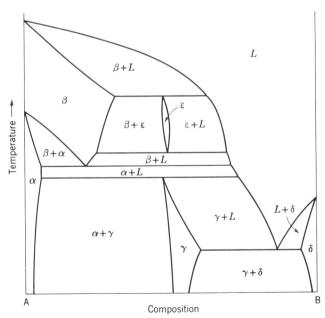

Fig. 8.3   *A corrected form of the diagram in Fig. 8.2.*

8.2   CALCULATION OF COMPLEX BINARY DIAGRAMS

The application of thermodynamic calculations to the determination of complex phase diagrams is not yet generally feasible. However, the science of phase diagrams has been proceeding in this direction, and some significant success has been possible with diagrams of intermediate complexity. As an example of such calculations we shall examine the work of Kaufman[8.1] on the Fe-Ru system.

The experimental Fe-Ru system is presented in Fig. 8.4a as given by Kaufman based on experimental data of other investigators. It may be seen that this is a system involving three invariant reactions, two high-temperature peritectics and a eutectoid at lower temperatures. The phases involved are a liquid solution and four solid solutions, the latter based, respectively, on the α (bcc), γ (fcc), and δ (bcc) polymorphic forms of iron and the single solid form of ruthenium, a hexagonal close-packed (hcp) phase.

To arrive at free-energy equations for these liquid and solid solutions, Kaufman used expressions of the form of Eqs. (4.37) and (4.38). He assumed that all solutions could be treated as regular solutions ($\Delta S^{\mathrm{xs}} = 0$) and that at constant temperature the excess free energies varied symmetrically and parabolically with composition, as in Eq. (4.47). On the basis of these considerations and some measurable physical properties of pure iron and ruthenium (in particular, the magnetic properties, the transformation temperatures, and the specific heats), he was able to calculate equations for the free energies of the various phases as a function of temperature and composition. Applying the common-tangent criterion as expressed in Eq. (6.4) he then derived equations giving the equilibrium solubilities of any two phases in the system as a function of temperature. Two of these equations, those for the boundaries of the $\gamma + \varepsilon$ field in the Fe-Ru diagram, are given below:

$$\Delta G_{\mathrm{Fe}}{}^{\gamma \to \varepsilon} + RT \ln \frac{1 - X_{\mathrm{Ru}}{}^{\varepsilon}}{1 - X_{\mathrm{Ru}}{}^{\gamma}} = 3800(X_{\mathrm{Ru}}{}^{\varepsilon})^2 - 2000(X_{\mathrm{Ru}}{}^{\gamma})^2$$

$$\tag{8.1}$$

$$\Delta G_{\mathrm{Ru}}{}^{\gamma \to \varepsilon} + RT \ln \frac{X_{\mathrm{Ru}}{}^{\varepsilon}}{X_{\mathrm{Ru}}{}^{\gamma}}$$
$$= 3800(1 - X_{\mathrm{Ru}}{}^{\varepsilon})^2 - 2000(1 - X_{\mathrm{Ru}}{}^{\gamma})^2 \tag{8.2}$$

In these equations, $\Delta G_{\mathrm{Fe}}{}^{\gamma \to \varepsilon}$ is the free-energy change accompanying

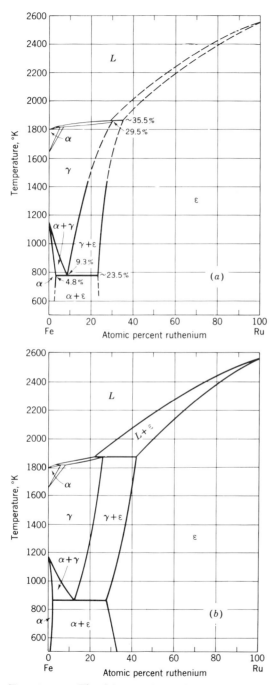

Fig. 8.4 *The (a) experimental and (b) calculated iron-ruthenium equilibrium phase diagram. (From Kaufman.[8.1])*

the transformation of pure iron from the γ to the ε form at the temperature in question, and similarly $\Delta G_{Ru}^{\gamma \to \varepsilon}$ for ruthenium. Values for these functions were also calculated from magnetic and specific-heat data (listed in Table 8.1). Equations (8.1) and (8.2)

Table 8.1   Difference in Free
               Energy between the
               fcc and hcp Forms of
               Iron and
               Ruthenium[8.1]

T, °K	$\Delta G_{Fe}^{\gamma \to \varepsilon}$, cal/mole	$\Delta G_{Ru}^{\gamma \to \varepsilon}$, cal/mole
0	−153	−153
100	−165	−175
200	−140	−230
300	−75	−305
400	+20	−385
500	+115	−465
600	+235	−550
700	+340	−640
800	+445	−725
900	+565	−815
1000	+685	−905
1100	+805	−990
1200	+920	−1070
1300	+1035	−1150
1400	+1140	−1235
1500	+1240	−1320
1600	+1340	−1405
1700	+1430	−1485
1800	+1520	−1570

are the analogues of Eqs. (4.31), (6.11), and (6.12), where now the terms on the right of (8.1) and (8.2) appear because neither the assumption of ideality ($\Delta G^{xs} = 0$), as in deriving (4.31), nor of dilute solution ($X \ll 1$), as in deriving (6.11) and (6.12), has been made. Simultaneous solution of Eqs. (8.1) and (8.2) yields the phase boundaries of the $\gamma + \varepsilon$ field in the Fe-Ru diagram.

The entire Fe-Ru phase diagram calculated on this basis is shown in Fig. 8.4b. In view of the simplifying assumptions made in computing the diagram, the level of agreement between the calculated and experimental diagrams is quite good. The major differences may be seen to be in the widths of the $\varepsilon + \gamma$ and $\varepsilon + L$ regions and in the temperature of the $\gamma \rightarrow \alpha + \varepsilon$ eutectoid invariant.

8.3   THE EFFECT OF HIGH PRESSURE ON TWO-COMPONENT SYSTEMS

In view of Eqs. (3.7) and (3.8)

$$\left(\frac{\partial G}{\partial P}\right)_T = V \tag{3.7}$$

and

$$\left(\frac{\partial G}{\partial T}\right)_P = -S \tag{3.8}$$

it may be seen that the volume of a phase plays the same role in determining the free-energy change of a phase attendant on a change of pressure as does the entropy for a change in temperature. This is also apparent from the definitional equation for free energy

$$G \equiv E + PV - TS \tag{2.9}$$

However, as pointed out in Chaps. 3 and 4, in solids and liquids at ordinary pressures the PV term is negligible compared to the terms E and TS, but it becomes the same order of magnitude as these terms at pressures of tens or hundreds of kilobars. At these high pressures, therefore, the PV term must be included in calculating, or estimating, the free energy. Furthermore, since in general the volumes of different phases are different, the magnitudes of the PV terms for the various phases in a system will not be the same, and, as a result the relative stabilities of the phases may change significantly as the pressure increases. Again as pointed out in Chap. 3, the free energy always increases as the pressure is raised (at constant temperature and composition, see Fig. 3.7) since both the pressure and volume are positive. Thus, a phase or phase mixture with low volume is favored by an increase in pressure, and the region of stability of such a phase or phase mixture in a phase diagram is extended at the expense of higher-volume phases.

It should be clear from Eq. (2.9) that a detailed analysis of the manner in which the PV term affects phase stability could be made if the volume could be expressed as a function of composition, pressure, and temperature. This can as yet only be done empirically; the volume of a solution, just like the enthalpy, entropy, and free energy, is usually written

$$V^S = V^M + \Delta V^{xs}$$

where V^M is the volume of the mechanical mixture of the components and ΔV^{xs} is the excess volume of solution, usually called simply the mixing volume. V^M is, of course, a linear function of X at a given temperature and pressure, but ΔV^{xs} is not. Fortunately, however, in many systems ΔV^{xs} is small and/or varies in a simple manner with composition. Examples are given in Fig. 8.5. In addition, V is expected to vary only slowly and nearly linearly with both P and T over wide ranges of P and T for many solids and liquids since both the compressibility and the thermal-expansion coefficient are small and nearly constant. Thus, in a qualitative or semiquantitative analysis of the effect of the PV term on equilibria in condensed

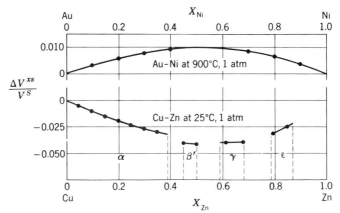

Fig. 8.5 $\Delta V^{xs}/V^S$ *for Au–Ni solid solutions at 900°C and 1 atm pressure and for Cu–Zn solid solutions at 25°C and 1 atm pressure. (Data from Kubaschewski and Catterall.[8.2])*

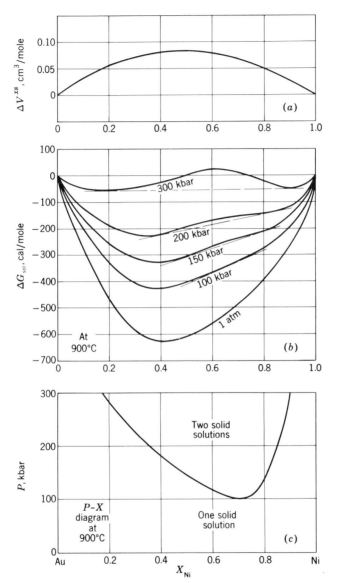

Fig. 8.6 ΔV^{xs} *and* ΔG_{sol} *for Au–Ni solid solutions at 900°C and 1 atm pressure, and the resulting* ΔG_{sol} *curves and pressure-composition phase diagram at 900°C.*

systems it is sufficient for many purposes to neglect the effect of
pressure and temperature on the volumes and on ΔV^{xs} and consider only the change in the product $P \, \Delta V^{xs}$.

In this vein, let us consider the Au–Ni phase diagram, shown
at 1 atm pressure in Fig. 4.23. In the solid state, this system exhibits
complete intersolution at high temperatures and a miscibility gap
at low temperatures. Solution-free-energy data measured at 900
and 700°C are found to be consistent with this diagram (Fig. 4.23).
Data for the mixing volume for this system at room temperature
are given in Fig. 8.6a, calculated from the curve in Fig. 8.5 and the
molal volumes of gold and nickel. If we now make the approximation
that the curve of ΔV^{xs} versus composition does not change with
temperature and pressure, we may calculate the effect of high
pressures on the phase diagram at, say, 900°C. To do this we
calculate $P \, \Delta V^{xs}$ as a function of composition at several different
pressures and add the resulting curves to the ΔG_{sol}-X curve at
1 atm, giving the curves shown in Fig 8.6b. It is seen that, because
of the positive ΔV^{xs}, at sufficiently high pressures the free-energy
curves develop an inflection. This leads to a region of two-phase
stability, a miscibility gap, at 900°C, just as lowering the temperature
does at 1 atm pressure. This is because the phase mixture of Au-rich
and Ni-rich phases has lower volume than a single solution of
intermediate composition and, thus, is favored by the high pressure;
the system relieves the constraint of high pressure by forming
the low-volume phase mixture (the Le Chatelier principle). The
resulting P-X phase diagram at 900°C is given in Fig. 8.6c.

In studying the effect of pressure on specific systems, it is generally more useful, and thus more common, to examine the effect of
pressure on the T-X phase diagram at several fixed high pressures
rather than the effect of pressure at some fixed temperature. The
number of ways in which high pressure can change T-X phase
diagrams is extremely large. A few illustrative schematic examples
are given in Figs. 8.7 and 8.8. In each figure typical free-energy–composition curves and the resulting phase diagram at atmospheric
pressure are given at the left, the effect of pressure in the columns
on the right. In Fig. 8.7, the diagram at atmospheric pressure is one
of complete mutual solubility in both the high-temperature and
low-temperature phases. If the components, A and B, are assumed
to mix ideally ($\Delta V^{xs} = 0$), as in Fig. 8.7a, high pressure simply
extends the stability range of the denser phase at the expense of
the less dense. If, as in Fig. 8.7b, the low-temperature phase has a

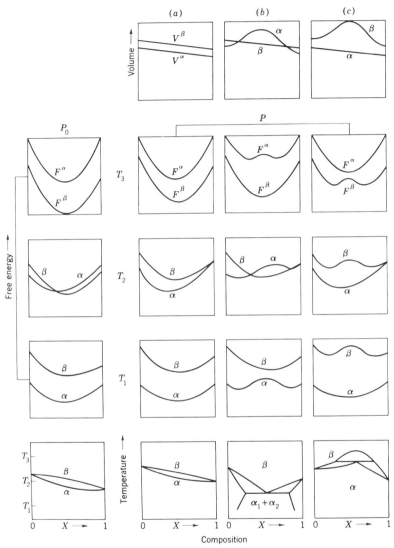

Fig. 8.7   *Schematic representation of the effects of pressure on two-phase equilibria. The free-energy curves and T-P phase diagrams are shown at atmospheric pressure P_0 and at high pressure P for the three different cases of relative volumes indicated in (a) to (c). (From Kaufman.*[8.3]*)*

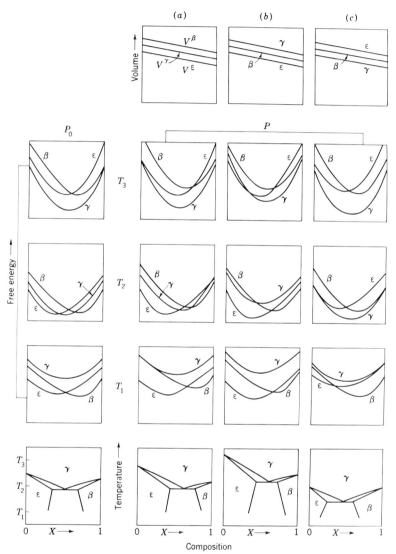

Fig. 8.8 *Schematic representation of the effects of pressure on three-phase equilibria. The free-energy curves and T-P phase diagram are shown at atmospheric pressure (P_0) and at high pressure (P) for the three different cases of relative volumes indicated in (a) to (c). (From Kaufman.*[8.3]*)*

positive ΔV^{xs}, high pressure produces a eutectic-like three-phase invariant reaction in the diagram; if, on the other hand, the high-temperature phase has a positive ΔV^{xs} (Fig. 8.7c), a peritectic-like invariant (in this case one with the form of the syntectic diagram) appears at high pressures. In Fig. 8.8, it is assumed that the atmospheric-pressure diagram is of the simple eutectic (or eutectoid) form and that all three phases concerned have $\Delta V^{xs} = 0$. In this case, the form of the diagram is not altered at high pressure, but the various phase regions are expanded or contracted depending

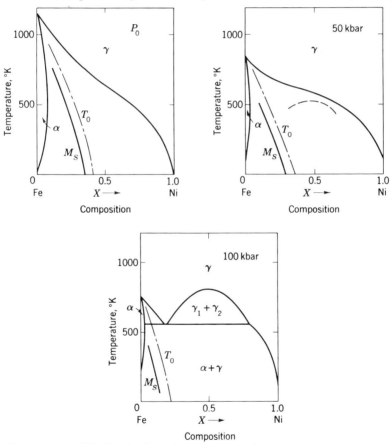

Fig. 8.9   *Calculated effect of pressure on the α-γ equilibria in the iron-nickel system. Experimental curve at atmospheric pressure (top) and calculated curves for pressures of 50 and 100 kbar. (From Kaufman and Ringwood.[8.4])*

on whether the corresponding phases have relatively low or relatively high volumes.

An illustration of the predicted appearance of a eutectoid invariant in a real system at high pressure is given for the Fe–Ni system in Fig. 8.9, adapted from the work of Kaufman and Ringwood.[8.4] These calculations were carried out to support the suggestion by the authors that a high-pressure euctectoid reaction may be the explanation for the observed appearance of a pearlitelike microstructure in metallic meterorites.

8.4   MULTICOMPONENT SYSTEMS

Detailed discussion of ternary and other multicomponent systems is beyond the intended scope of this book. Several excellent treatments from the geometrical (Refs. 6.1, 8.5, and 8.6) and the topological (Ref. 8.7) viewpoints are available in the literature. Little has been done, however, along thermodynamic lines, for the associated equations are of forbidding length, and the scarcity of data drastically limits their prospective utility. The discussion here will therefore be confined to a few qualitative illustrations for the simplest ternary systems.

Ternary systems involve four variables, the temperature, the pressure, and two composition variables. The complete phase diagrams of such systems are therefore four-dimensional and cannot be drawn. These diagrams are generally handled by fixing one variable, and, for condensed systems, this is usually the pressure. In such constant-pressure three-dimensional sections, the temperature is customarily plotted vertically and the two independent composition variables horizontally. As a result, the meaningful portion of the horizontal composition plane is, perforce, a triangular region, e.g., that between the points A, B, and C in Fig. 8.10. This may be demonstrated by noting that, as the axes are labeled in Fig. 8.10a, the point P represents an alloy containing no A, since in it the mole fractions of B and C are both 0.5. Similarly, all alloys containing no A lie along the line CPB, and the A content of alloys increases in the direction PA. Thus, all possible alloy compositions lie within the triangle ABC, and the portion of the composition plane CMB has no physical significance.

It is clear from an inspection of Fig. 8.10a that the three vertical sides of the constant-pressure phase diagram are simply the

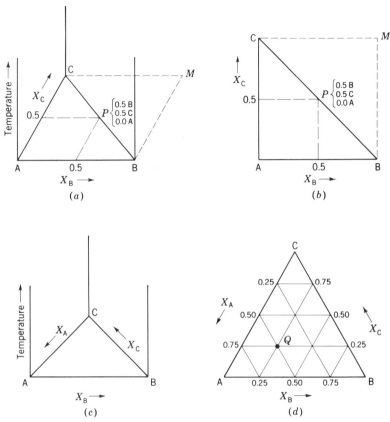

Fig. 8.10 *The composition plane in ternary (three-component) diagrams.*

three binary diagrams making up the ternary system. However, as drawn in Fig. 8.10a and b, with the two independent composition axes orthogonal to each other (the usual method in other types of graphical representations) it is found that the composition axis for one of the three components differs in length from the other two. To remove this inconvenience it has become customary to represent the composition plane as an equilateral triangle, as in Fig. 8.10c and d. Compositions are then read as indicated in Fig. 8.10d; e.g., the point Q represents an alloy with mole fractions 0.50 A, 0.25 B, and 0.25 C.

The problem of determining the equilibrium states in three-component systems is again, just as for one-component and two-component systems, one of determining the lowest-free-energy conditions at constant temperature and pressure on the basis of equations of the type

$$G = f(T,P,X_1,X_2) \tag{3.1}$$

for each possible phase in the system. The plots of constant-temperature constant-pressure free energy versus composition are now, however, surfaces rather than lines. Two simple schematic illustrations will be discussed here.

In the first illustration, let us suppose that three components A, B, and C are mutually soluble in each other in all proportions, both in the liquid and in the solid states. We then have the ternary counterpart of the binary diagrams shown in Fig. 4.6 or 4.32*b*. Such a diagram is illustrated in Fig. 8.11*a*. The liquidus and solidus have now become surfaces the shapes of which can be visualized by imagining that they are generated by carrying the binary liquidus and solidus lines into the ternary space. A constant-temperature (horizontal) plane passed through the diagram would, at an appropriate temperature, intersect these surfaces in lines such as *aa* and *bb* in the diagram. The line *aa* is the locus of composition points representing liquids which at this temperature are in equilibrium with solid solutions of compositions lying along the line *bb*. The two lines *aa* and *bb* are derived from the free-energy surfaces for the solid and the liquid at this temperature in the manner illustrated in Fig. 8.11*c*. The two free-energy surfaces, each of which has the shape of a three-cornered hammock, intersect each other at this temperature. If we imagine now that a plane is placed from below against these convex-downward free-energy surfaces, this plane will be tangent to the two surfaces at two points; these two points will, for example, be the points *a'* and *b'* in Fig. 8.11*c* if the tangent plane is appropriately placed, and by analogy with the two-component case, it may be seen that *a'* and *b'* are the equilibrium phase compositions for two-phase alloys in the binary system A-C. The tangent plane is the three-component counterpart to the common tangent line we have discussed for two-component systems, such as the line *a'b'* in the system A-C of Fig. 8.11*c*. In the three-component case, however, the tangent plane can "rock" along the convex free-energy surfaces while always remaining tangent to the two surfaces. In so rocking it generates an infinite number of pairs of common tangency points, the loci of which form the lines *a'a''* and *b'b''*, as in Fig. 8.11*c*. These lines projected onto a horizontal plane, say

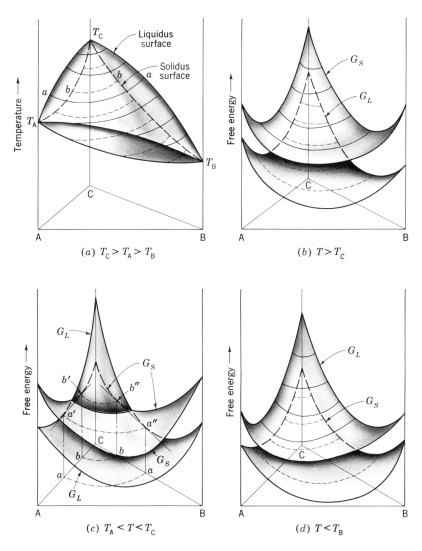

Fig. 8.11   *The ternary complete solid-solution dia-gram and corresponding free-energy surfaces at three representative temperatures.*

the basal plane in Fig. 8.11c, become the lines *aa* and *bb*, the equilibrium solubility lines shown in Fig. 8.11a. In the three-component system, for any alloy which at the temperature in question contains both solid and liquid in equilibrium, the compositions of the two phases will be given by one of the pairs of common tangency points generated by the rocking plane; the particular pair of tangency points will be that for which a straight line between them runs through the alloy composition point.

For this same ternary alloy system, at temperatures above the melting point T_c of the highest-melting pure component the liquid free-energy surface will be lower than that of the solid at all compositions, as in Fig. 8.11b. In this case all alloys have their lowest-free-energy, i.e., equilibrium, condition if they are completely liquid, as indicated in the phase diagram. The reverse is true at temperatures below the melting point T_B of the lowest-melting pure component. Here the free energy of the solid is lower than that of the liquid at all alloy compositions (Fig. 8.11d), and hence the solid is stable at all compositions.

As a second example, let us consider a ternary system in which the three components are mutually soluble to only a limited extent in the solid state, and in which no intermediate phases appear. The partial phase diagram of such a system at temperatures in the completely solid range is presented in Fig. 8.12a. Alloys in this system which have overall compositions lying within the triangle abc at the temperature T_1 will contain three solid solutions in equilibrium, and the compositions of the three phases will be, respectively, a, b, and c. The three equilibrium compositions are derived from the free-energy surfaces at this temperature, as shown in Fig. 8.12b. For simplicity in drawing and reading the latter figure, the free-energy axis has been inverted, so that the free-energy surfaces are now convex upward rather than downward. It may be seen by analogy again that the lowest-free-energy condition can be ascertained by placing a plane so that it is tangent to all three free-energy surfaces. The points of common tangency, a', b', and c', give the equilibrium compositions a, b, and c of the three solid solutions. It should be clear that now there is no possibility of rocking the tangent plane; the points of common tangency (and, thus, the compositions in equilibrium) are unique for each such temperature. This means that for all alloys in which these three, and only these three, phases coexist at equilibrium, the phase compositions will be the uniquely determined values a, b, and c. This, it may be noted, is consistent with the predictions of the phase rule; we have three components and three phases, and, hence, according to the phase rule,

$$f = 3 - 3 + 2 = 2$$

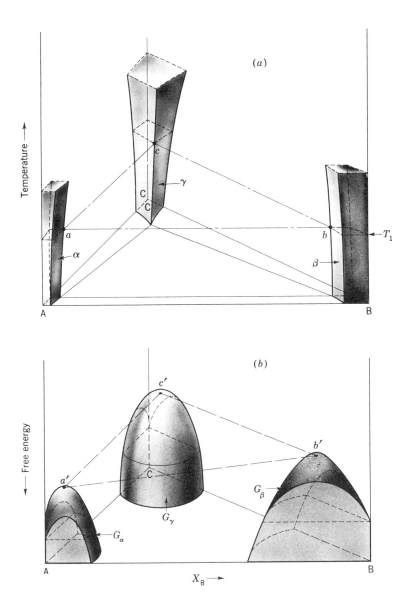

Fig. 8.12 *Partial phase diagram (a) and corresponding free-energy surfaces (b) at a representative temperature. (Free-energy axis is inverted from the usual
direction.)*

The two available degrees of freedom have been utilized in arbitrarily fixing the pressure and the temperature; hence, all phase compositions are uniquely fixed by nature. In the case of the *two*-phase equilibria discussed in the first three-component system, there was one more degree of freedom; thus the composition of one of the phases could be selected arbitrarily in addition to the temperature and the pressure. This was equivalent to selecting the specific position of the rocking tangency plane; but once this was done, the composition of the other phase was fixed by nature; i.e., the selection of one of the phase compositions fixed the position of one common tangency point, and this in turn automatically fixed the position of the second tangency point in the pair selected.

Problems

8.1 Examine the Ag–Zn phase diagram (cf. Ref 4.17). Redraw on an expanded scale that portion of the diagram lying between 200 and 300°C and 20 to 50 weight percent Zn.

8.2 (a) What is the value of p_{max} in a three-component equilibrium diagram? (b) Construct a table showing the dimensionality of the boundaries between all the possible pairs of p-phase regions in such a diagram.

8.3 Repeat Prob. 8.2 assuming the pressure is arbitrarily fixed.

8.4 Draw a single complex two-component diagram (pressure fixed) containing one each of four different types of invariant reactions.

8.5 List the invariant reactions in the Cu–Si (cf. Ref 4.17) phase diagram, giving for each the name, temperature, and reaction, including the compositions of the phases taking part.

8.6 Redraw the diagram in Fig. 8.3 eliminating the ε phase but otherwise making as few changes as necessary to keep the diagram theoretically correct.

8.7 From Eqs. (8.1) and (8.2) and the data in Table 8.1 calculate the equilibrium solubilities of the γ and ε phases in the Fe–Ru phase diagram at the peritectic (1880°K) and at the eutectoid (780°K) temperatures. Compare with Fig. 8.4.

8.8 From the data in Figs. 4.23 (bottom) and 8.6a calculate the free-energy–composition curve for the Au–Ni system at 900°C and 500 kbar. (Make the same assumptions made in the text on page 198.) Plot these data carefully and determine the intersolubility limits under these conditions.

8.9 In a fashion similar to that shown in Fig. 8.8, construct schematic free-energy curves and the resulting phase diagrams at atmospheric pressure and at some very high pressure for a simple peritectic diagram, assuming that V^L, V^α, and V^β vary linearly with composition and that (a) $V^L > V^\alpha > V^\beta$ and (b) $V^\alpha > V^\beta > V^L$.

References

8.1 L. Kaufman, Thermodynamics of Martensitic FCC-BCC and FCC-HCP Transformations in the Iron-Ruthenium System, *Proc. Conf. Phys. Met. Martensite Bainite Brit. Iron Steel Inst., Inst. Met.*, May, 1965, Scarborough, England.

8.2 O. Kubaschewski and J. A. Catterall, "Thermochemical Data of Alloys," Pergamon Press, New York, 1956.

8.3 L. Kaufman, Some Equilibria and Transformations in Metals under Pressure, in Paul and Warschauer (eds.), "Solids under Pressure," McGraw-Hill Book Company, New York, 1963.

8.4 L. Kaufman and A. E. Ringwood, *Acta Met.*, **9**:941 (1961).

8.5 G. Masing and B. A. Rogers, "Ternary Systems," Reinhold Publishing Corporation, New York, 1944.

8.6 A. Prince, "Alloy Phase Equilibria," American Elsevier Publishing Company, New York, 1966.

8.7 L. S. Palatnik and A. I. Landau," Phase Equilibria in Multicomponent Systems," Holt, Rinehart and Winston, Inc., New York, 1964.

SOLUTIONS TO SELECTED PROBLEMS

CHAPTER 2

2.2 Consider two arbitrarily chosen paths a and b both of which carry a system from state A to state B. If $\Delta E_{AB}{}^a$ and $\Delta E_{BA}{}^a$ are the internal energy changes in going along path a from state A to state B and from state B to state A, respectively, and $\Delta E_{AB}{}^b$ and $\Delta E_{BA}{}^b$ are the corresponding changes for path b, then the statement of the first law in the problem reveals that

$$\Delta E_{AB}{}^a + \Delta E_{BA}{}^b = 0$$

and

$$\Delta E_{AB}{}^b + \Delta E_{BA}{}^b = 0$$

Subtracting, we find that

$$\Delta E_{AB}{}^a = \Delta E_{AB}{}^b$$

Since paths a and b were arbitrarily chosen paths, a similar argument can be advanced for any other two paths; thus, ΔE_{AB} is independent of path.

2.7 Consider the interfacial surface between two contiguous phases α and β in which there are the two components A and B. Let $n_A{}^\alpha$ and $n_A{}^\beta$ be the number of A atoms per unit volume at the interface on the α side and on the β side of the interface, respectively, and similarly let $p_{AB}{}^{\alpha\to\beta}$ be the probability that an A atom in phase α will change places with a B atom in phase β, and $p_{AB}{}^{\beta\to\alpha}$ be the corresponding probability for an A atom in phase β and a B atom in phase α. Then the flux of A atoms from α to β is

$$\text{flux}_A{}^{\alpha\to\beta} = n_A{}^\alpha p_{AB}{}^{\alpha\to\beta}$$

and the flux of A atoms from β to α is

$$\text{flux}_A{}^{\beta\to\alpha} = n_A{}^\beta p_{AB}{}^{\beta\to\alpha}$$

If equilibrium is to be maintained, there can be no composition changes, and therefore no *net* flux, i.e.,

$$n_A{}^\alpha p_{AB}{}^{\alpha\to\beta} = n_A{}^\beta p_{AB}{}^{\beta\to\alpha}$$

The probabilities p are functions of the crystal structures of the phases or, more precisely, of the numbers and kinds of interatomic bonds about each atom in the phases. Since, in general, these are different for different phases then

$$p_{AB}{}^{\alpha \to \beta} \neq p_{AB}{}^{\beta \to \alpha}$$

and therefore

$$n_A{}^\alpha \neq n_A{}^\beta$$

CHAPTER 3

3.1 We wish to find $S_{1073°C} = S_{1346°K}$. From Eq. (3.5)

$$dS = \frac{C_P}{T} dT$$

Therefore

$$\Delta S = S_{1346} - S_{300} = \int_{300}^{1346} \frac{5.41 + 1.50 \times 10^{-3}T}{T} dT$$

$$= 9.68 \text{ cal/(mole)}(°K)$$

$$S_{1346} = S_{300} + 9.68 = 17.68 \text{ cal/(mole)}(°K)$$

(For entropy data see, for example, J. F. Elliott and M. Gleiser, "Thermochemistry for Steelmaking," Vol. I, Addison-Wesley Publishing Co., Inc., Cambridge, Mass., 1960.)

3.3 From Eq. (2.10), the change in free energy of the copper is

$$dG = V \, dP - S \, dT$$

Then, we must show that, for the copper block,

$$\frac{\int_{P=1}^{P=P_{\text{final}}} V \, dP}{\int_{300}^{1000} S \, dT} < 0.01$$

To find the final pressure in the chamber, we may assume the gas behaves ideally if the pressure rise is not more than a few atmospheres. Thus, for the gas,

$$P_1V_1 = RT_1$$
$$P_2V_2 = RT_2$$

and, since $V_1 = V_2$, then

$$\frac{P_2}{P_1} = \frac{T_2}{T_1} = \frac{1000}{300} = \frac{10}{3}$$

Since solids are highly incompressible, we may assume V of the copper will not change for this relatively small rise in pressure, so that

$$\int V \, dP = V \, \Delta P$$

The molal volume of copper is about 7 cm³, and therefore, per mole of copper,

$$V \, \Delta P \cong 7 \times \tfrac{10}{3} \, (\text{cm}^3)(\text{atm}) \cong 0.57 \, \text{cal/mole}$$

From the equations in Prob. 3.1

$$\int_{300}^{1000} S \, dT = \int\!\!\int_{300}^{1000} \frac{C_P}{T} \, dT \, dT = \int\!\!\int_{300}^{1000} \frac{5.41 + 1.50 \times 10^{-3T}}{T} \, dT \, dT$$
$$= 9950 \, \text{cal/mole}$$

Thus

$$\frac{\int V \, dP}{\int S \, dT} = \frac{0.57}{9950} \cong 0.57 \times 10^{-4}$$

and the change in free energy due to pressure change may be neglected.

3.4   As a sample calculation we shall calculate the melting temperature of iron at 10^5 atm. From Eq. (3.17)

$$\frac{dT}{dP} = \frac{\Delta V}{\Delta H} T$$

which upon integration yields

$$\ln T - \ln T_0 = \frac{\Delta V}{\Delta H} (P - P_0)$$

At $P = 1$ atm, $T = T_0 = 1809°K$, and $\Delta V = 0.231$ cm³/mole, $\Delta H = 3700$ cal/mole. Thus

$$\ln \frac{T}{1809} = \frac{0.231}{(3700)(41)} P = 1.52 \times 10^{-6} P$$

$$\log \frac{T}{1809} = 6.72 \times 10^{-7} P$$

For $P = 10^5$ this gives $T = 2110°K = 1837°C$.

3.5 The stacking-fault energy is the difference in free energy between the fcc and the bcc forms of iron. The difference in enthalpy ΔH for the two forms can be calculated roughly from the slope of the γ-ϵ line in the phase diagram and Eq. (3.17). ΔS can be estimated as $\Delta H/T_m$. This then gives ΔG, the fault energy per unit volume of stacking fault.

3.7   From Eq. (3.10)

$$\beta \equiv -\frac{1}{V}\frac{dV}{dP} \quad \text{or} \quad \frac{dV}{V} = -\beta \, dP \quad \text{at constant } T$$

We wish to find the P at which $V^\alpha = V^\gamma$.
   Integrating the above

$$\ln \frac{V_2}{V_1} = -\beta(P_2 - P_1) = -\beta P_2 \quad \text{since } P_2 \gg P_1 = 1$$

Thus

$$\ln \frac{V_2{}^\alpha}{V_1{}^\alpha} = -\beta^\alpha P$$

and

$$\ln \frac{V_2{}^\gamma}{V_1{}^\gamma} = -\beta^\gamma P$$

Subtracting these two equalities

$$\ln \frac{V_2{}^\alpha V_1{}^\gamma}{V_1{}^\alpha V_2{}^\gamma} = P(\beta^\gamma - \beta^\alpha)$$

But

$$\frac{V_1{}^\alpha}{V_1{}^\gamma} = 1.02 \qquad \text{and} \qquad \beta^\alpha = 1.5\beta^\gamma$$

Thus, for $V_2{}^\alpha = V_2{}^\gamma$

$$\ln 1.02 = \tfrac{1}{3}\beta_\alpha P$$

$$P = \frac{3(2.3 \log 1.02)}{5.9 \times 10^{-7}} \cong 1 \times 10^5 \text{ atm}$$

3.9   Since

$$\frac{\partial^2 G}{\partial T^2} = -\frac{C_P}{T} \tag{3.12}$$

then

$$\Delta\left(\frac{\partial^2 G}{\partial T^2}\right)_{T_E} = -\left(\frac{\Delta C_P}{T}\right)_{T_E}$$

In a second-degree transformation this is not zero; therefore $(\Delta C_P)_{T_E} \neq 0$ at finite temperatures.

In a third-degree transition, this is zero; therefore $(\Delta C_P)_{T_E} = 0$ at finite temperatures.

CHAPTER 4

4.1   Consider two ideal gases, A and B, in a single adiabatic, isometric container, but initially separated by an impermeable membrane. If the membrane is removed, the gases will come to a uniform mixture. Since in an ideal gas there are no interaction energies between particles, the potential energy of the system does not change, and, since no heat can flow through the container walls, the kinetic energies of the particles do not change. Thus, both E and T of the system remain constant. There is a change in state, however, due to the mixing, and an entropy change of mixing. This entropy change can be found by ascertaining the individual changes in entropy of each gas and adding them, for, as far as each gas is concerned, it simply went through a constant-temperature, adiabatic expansion. If this change for each gas were carried out reversibly, the first law for the change could be written

$$dE = T\, dS - P\, dV = 0$$

or

$$dS = \frac{P\, dV}{T}$$

and, from the ideal gas law, $P/T = nR/V$, so that

$$dS = nR\, \frac{dV}{V}$$

Integrating,

$$\Delta S = nR \ln \frac{V^f}{V^i}$$

where V^f = final volume of gas
V^i = initial volume of gas

For the two gases together, then

$$\Delta S_A + \Delta S_B = \Delta S_m = n_A R \ln \frac{V_A{}^f}{V_A{}^i} + n_B R \ln \frac{V_B{}^f}{V_B{}^i}$$

If we now take $n = n_A + n_B = 1$, then

$$n_A = X_A \qquad n_B = X_B$$

and

$$V_A{}^i = X_A V \qquad V_B{}^i = X_B V$$

$$V_A{}^f = V_B{}^f = V = \text{total volume of container}$$

Hence

$$\Delta S_m = X_A R \ln \frac{V}{X_A V} + X_B R \ln \frac{V}{X_B V}$$

$$\Delta S_m = -R(X_A \ln X_A + X_B \ln X_B)$$

4.2 Let n = number of atomic diameters in the linear dimension of the particle

E = energy/volume in bulk

$2E$ = energy/volume in surface

V = volume of particle

V_S = volume of the surface of particle

We wish to know when

$$\frac{2EV_S}{EV} = 0.1$$

or when

$$\frac{V_S}{V} = 0.05$$

The number of atoms in $V = Kn^3$ and in $V_S = 2K'n^2$, where K and K' are shape constants.

If we now assume that the number of atoms in a volume is directly proportional to the volume, then $V_S/V = 0.05$ when

$$\frac{V_S}{V} = \frac{2K'n^2}{Kn^3} = 0.05$$

or

$$n = 40\,\frac{K'}{K}$$

If the particles are spheres, $K' = \pi$ and $K = \dfrac{\pi}{6}$ so that

$$n = 240 \text{ atomic diameters}$$

4.7 Note that

$$\frac{d^L}{d^S} = \frac{2\sqrt{D^L t}}{2\sqrt{D^S t}}$$

For equal penetration times

$$\frac{d^L}{d^S} = \sqrt{\frac{D^L}{D^S}} = \frac{9.15 \times 10^{-4}\exp\left(-\dfrac{4450}{RT}\right)}{0.28\exp\left(-\dfrac{24{,}200}{RT}\right)}$$

At the melting point of lead ($900°K$), from the above

$$\frac{d^L}{d^S} = 5.72 \times 10^{2.4} \cong 1.4 \times 10^3$$

4.9   From Eq. (4.51)

$$NZ\mathcal{U} = 2RT_c = 2R(1260 + 273) = 6130 \text{ cal/mole}$$

From Eq. (4.47)

$$\Delta H^{xs} = NZ\mathcal{U}X_AX_B = (6130)(0.5)(0.5) = 1533 \text{ cal/mole}$$

(*a*) At $1200°C = 1473°K$,

$$T\,\Delta S_m = -RT(X_A \ln X_A + X_B \ln X_B)$$
$$= -(2)(1473)(\ln 0.5) = 2020 \text{ cal/mole}$$

(*b*) At $700°C = 973°K$,

$$T\,\Delta S_m = -(2)(973)(\ln 0.5) = 1330 \text{ cal/mole}$$

4.10   (*a*) Assuming regular solutions

$$\Delta G^{sol} = \Delta G^{xs} - T\,\Delta S_m = \Delta H^{xs} - T\,\Delta S_m$$
$$\Delta H^{xs} = \Delta G_{sol} + T\,\Delta S_m$$
$$T\,\Delta S_m = -RT(X_A \ln X_A + X_B \ln X_B)$$
$$= -(4.6)(1173)(0.7 \log 0.7 + 0.3 \log 0.3)$$
$$= 1420 \text{ cal/mole}$$

From the free-energy curves, $\Delta G^{sol} = -490$ cal/mole

$$\Delta H^{xs} = -490 + 1420 = 930 \text{ cal/mole}$$

(*b*) From (4.47) and (4.50)

$$\Delta H^{xs} = NZ\mathcal{U}X_AX_B = 2RT_cX_AX_B = 4(812 + 173)(0.7)(0.3)$$
$$= 910 \text{ cal/mole}$$

(*c*)   The values are close enough to indicate the solutions are approximately regular, provided the assumptions involved in calculating $\mathcal{U}$ in (*b*) are valid (see text).

CHAPTER 5

5.2 (a) The nearest neighbors of, say, the body-centered atom are the corner atoms; there are 8 of these. The second-nearest neighbors of, say, a corner atom are the other nearest corner atoms; there are 6 of these. The third-nearest neighbors of, say, a corner atom are the nearest corner atoms at the ends of the face diagonals; there are 12 of these.

(b) Calling the unit-cell edge length a, the nearest-neighbor distance d_1 is half the body diagonal, which is

$$d_1 = \frac{1}{2} \sqrt{3a^2} = a \frac{\sqrt{3}}{2}$$

The second nearest-neighbor distance is the unit-cell edge length

$$d_2 = a$$

The third-nearest-neighbor distance is the face diagonal

$$d_3 = \sqrt{2a^2} = a \sqrt{2}$$

5.3 If the total energy contributed by the pairs around each atom is E

$$E = n_1 v_1 + n_2 v_2 + n_3 v_3$$

where n_1, n_2, and n_3 are the number of first-, second-, and third-nearest neighbors, respectively, and v_1, v_2, and v_3 are the first-, second-, and third-nearest-neighbor interaction energies, respectively.

From Prob. 5.2

$$d_2 = \frac{2}{\sqrt{3}} d_1$$

$$d_3 = \frac{2 \sqrt{2}}{\sqrt{3}} d_1$$

Thus

$$v_2 = \left(\frac{\sqrt{3}}{2} \right)^6 v_1 = 0.419 v_1$$

$$v_3 = (\sqrt{\tfrac{2}{3}})^6 v_1 = 0.0189 v_1$$

There are 8 first-, 6 second-, and 12 third-nearest neighbors for each atom. Therefore

$$E = 8v_1 + 6(0.419v_1) + 12(0.0189v_1)$$

Percent of E for first neighbors $= \dfrac{8}{10.75}\,(100) = 74.4\%$

Percent of E for second neighbors $= \dfrac{6(0.419)}{10.75}\,(100) = 23.4\%$

Percent of E for third neighbors $= \dfrac{12(0.0189)}{10.75}\,(100) = 2.2\%$

5.6 Consider first the terms in Eq. (5.19) which have the coefficient $(-ZN/2)$. Substituting $\sigma = \mathcal{S}^2$ into each of these terms, we find

$$X_A{}^2(1 - \sigma) \qquad = X_A{}^2(1 - \mathcal{S}^2) \qquad = X_A(1 + \mathcal{S})X_A(1 - \mathcal{S})$$

$$X_B + \mathcal{S} + X_A\sigma = X_B + \mathcal{S} + X_A\mathcal{S}^2 = (1 + \mathcal{S})(X_B + X_A\mathcal{S})$$

$$X_B - \mathcal{S} + X_A\sigma \qquad\qquad = (1 - \mathcal{S})(X_B - X_A\mathcal{S})$$

$$X_B{}^2 - X_A{}^2\sigma \qquad\qquad = (X_B + X_A\mathcal{S})(X_B - X_A\mathcal{S})$$

Inserting these equalities and collecting terms gives for the terms having the coefficient $(-ZN/2)$

$$[X_A(1 + \mathcal{S})X_A(1 - \mathcal{S}) + X_A(1 + \mathcal{S})(X_B + X_A\mathcal{S})]\ln X_A(1 + \mathcal{S})$$
$$+ [X_A(1 + \mathcal{S})X_A(1 - \mathcal{S}) + X_A(1 - \mathcal{S})(X_B - X_A\mathcal{S})]\ln X_A(1 - S)$$
$$+ [X_A(1 - \mathcal{S})(X_B - X_A\mathcal{S}) + (X_B + X_A\mathcal{S})(X_B - X_A\mathcal{S})]\ln (X_B - X_A\mathcal{S})$$
$$+ [X_A(1 + \mathcal{S})(X_B + X_A\mathcal{S}) + (X_B X_A\mathcal{S})(X_B - X_A\mathcal{S})]\ln (X_B + X_A\mathcal{S})$$

which reduces to

$$X_A(1 + \mathcal{S}) \ln X_A(1 + \mathcal{S}) + X_A(1 - \mathcal{S}) \ln X_A(1 - \mathcal{S})$$
$$+ [1 - X_A(1 - \mathcal{S})] \ln [1 - X_A(1 - \mathcal{S})]$$
$$+ [1 - X_A(1 + \mathcal{S})] \ln [1 - X_A(1 + \mathcal{S})]$$

It will be noted this is identical with that part of Eq. (5.19) which has the coefficient $[(Z - 1)/2]N$. Thus, adding the two leaves the last expression above with the coefficient $(-N/2)$ as the expression for $\ln W$ corresponding to the Bragg-Williams condition.

5.8   Using Eq. (5.25) and the T_c-vs-composition data in Fig. 4.35 in the range 30 to 90 atomic percent, we find

Atomic percent Pt	$-V$, cal/mole
30	334
40	378
50	396
60	410
70	447
80	570
90	1330

It is seen that in the range 40 to 70 atomic percent Pt, V is reasonably constant. The large variation at the extremities of the composition scale is probably due both to progressively greater failure of the theory and the difficulty of obtaining equilibrium experimentally at the lower temperatures.

CHAPTER 6

6.2   (*a*) Using Eq. (6.2)

$$\frac{X_B{}^S}{X_B{}^L} = 1 + \frac{\Delta H_A{}^m}{RT_A{}^2}\frac{\Delta T}{X_B{}^L} \cong 0 \qquad \text{for} \qquad X_B{}^S \ll X_B{}^L$$

Thus, the slope, $\Delta T/X_B{}^L$, is approximately independent of the nature of the component B (the solute)

$$\frac{\Delta T}{X_B{}^L} = \text{slope} = \frac{RT_A{}^2}{RT_A{}^2 + \Delta H_A{}^m}$$

(*b*) Rearranging Eq. (6.2)

$$\frac{\Delta T}{X_B{}^L} = \text{slope} = \left(\frac{RT_A{}^2}{RT_A{}^2 + \Delta H_A{}^m}\right)\frac{X_B{}^S}{X_B{}^L}$$

All quantities on the right are constant except $\dfrac{X_B{}^S}{X_B{}^L}$; hence, slope is

proportional to $\dfrac{X_B{}^S}{X_B{}^L}$

6.5 From Eq. (6.3)

$$X_{Cu} = \exp\left(\frac{\Delta S^{xs}}{R}\right) \exp\left(-\frac{\Delta H^{xs}}{RT}\right)$$

Making the simplifying assumptions that the variations of ΔS^{xs} and ΔH^{xs} with T are negligible (justified by data such as that in Fig. 6.7), then

$$\ln \frac{X_{Cu}^{500°C}}{X_{Cu}^{200°C}} = \ln \frac{1.7}{0.68} = -\frac{\Delta H^{xs}}{R} \frac{(-773 + 473)}{(773)(473)}$$

$$\Delta H^{xs} = +2220$$

Thus,

$$\ln \frac{X_{Cu}^{400°C}}{X_{Cu}^{200°C}} = -\frac{2220}{R}\left(\frac{473 - 673}{473 \times 673}\right) = \ln X_{Cu}^{400} - \ln (6.8 \times 10^{-3})$$

$$X_{Cu}^{400} = 1.33 \text{ atomic percent}$$

6.6 (a) $\quad 100 \dfrac{X° - a}{e - a}$

(b) $\quad 100 \dfrac{b - e}{b - a} \dfrac{X° - a}{e - a}$

(c) $\quad 100 \dfrac{a - c}{d - c} \dfrac{e - X°}{e - a}$

where $X° =$ composition of alloy 3.

CHAPTER 8

8.2 (a) From the phase rule

$$f_{min} = 0 = c - p_{max} + 2$$

$$p_{max} = 5 \quad \text{when} \quad c = 3$$

(*b*) Here $m = 4$, for the temperature, pressure, and the two independent composition variables. Thus, from the general boundary rule (see page 188),

Possible pairs of contiguous p-phase regions	Difference in number of phases in contiguous regions, $m - n$	Dimensionality of boundary, n
5-4, 4-3, 3-2, 2-1	1	3
5-3, 4-2, 3-1	2	2
5-2, 4-1	3	1
5-1	4	0

8.3 (*a*) $p_{\max} = c + 1 = 4$

(*b*) Here $m = 3$

Possible pairs of contiguous p-phase regions	Difference in number of phases in contiguous regions, $m - n$	Dimensionality of boundary, n
4-3, 3-2, 2-1	1	2
4-2, 3-1	2	1
4-1	3	0

8.7 At 1880°K, by extrapolation of the data in Table 8.1,

$$\Delta G_{Fe}^{\gamma \to \epsilon} = +1590$$

$$\Delta G_{Ru}^{\gamma \to \epsilon} = -1635$$

Therefore,

$$1590 + 4.575 \times 1880 \log \frac{1 - X_{Ru}^{\epsilon}}{1 - X_{Ru}^{\gamma}} = 3800(X_{Ru}^{\epsilon})^2 - 2000(X_{Ru}^{\gamma})^2$$

$$-1635 + 4.575 \times 1880 \log \frac{X_{Ru}^{\epsilon}}{X_{Ru}^{\gamma}} = 3800(1 - X_{Ru}^{\epsilon})^2 - 2000(1 - X_{Ru}^{\gamma})^2$$

Adding,

$$-45 + 8610 \log \frac{(1 - X_{Ru}^{\epsilon})X_{Ru}^{\epsilon}}{(1 - X_{Ru}^{\gamma})X_{Ru}^{\gamma}}$$

$$= 3800[1 - 2X_{Ru}^{\epsilon} + 2(X_{Ru}^{\epsilon})^2] - 2000[1 - 2X_{Ru}^{\epsilon} + 2(X_{Ru}^{\gamma})^2]$$

This must be solved by approximation. Trying $X_{Ru}^{\gamma} = 0.25$ and $X_{Ru}^{\epsilon} = 0.41$ gives $900 = 720$; trying $X_{Ru}^{\gamma} = 0.27$ and $X_{Ru}^{\epsilon} = 0.37$ gives $700 = 820$. Thus, correct values are approximately

$$X_{Ru}^{\gamma} = 0.26 \qquad X_{Ru}^{\epsilon} = 0.39$$

The experimentally determined values from Fig. 8.4a are

$$X_{Ru}^{\gamma} = 0.295 \qquad X_{Ru}^{\epsilon} = 0.355$$

INDEX